ACKNOWLEDGEMENTS

The publishers gratefully acknowledge the permissions granted to reproduce copyright material in this book. Every effort has been made to contact the holders of copyright material, but if any have been inadvertently overlooked, the Publisher will be pleased to make the necessary arrangements at the first opportunity.

Chapter 1
p12 Smith1972/Shutter... ...ck, pan_kung/Shutterstock; p13 Olga Popova/Shutterstock, Albert Russ/Shutterstock, HUA... ...ock, motorolka/Shutterstock, ANDREW LAMBERT PHOTOGRAPHY/ SCIENCE PHOTO LIBRAR... ...y Karandaev/Shutterstock; p18 brumhildich/Shutterstock, Yenyu Shih/ Shutterstock; p20 Ufuk ZIVANA/Shutterstock; p21 PROF. PETER FOWLER/SCIENCE PHOTO LIBRARY; p22 GRAHAM J. HILLS/SCIENCE PHOTO LIBRARY; p24 SSSCCC/Shutterstock; p32 Boris15/Shutterstock; p33 ppl/Shutterstock; p34 demarcomedia/Shutterstock, Jeffrey B. Banke/ Shutterstock, Triff/Shutterstock, Gail Johnson/Shutterstock; p35 Anneka/Shutterstock, Mona Makela/Shutterstock, Asmus Koefoed/ Shutterstock, Popartic/Shutterstock; p40 DenisNata/Shutterstock; p41 tonyz20/Shutterstock; p42 ANDREW LAMBERT PHOTOGRAPHY/ SCIENCE PHOTO LIBRARY, tarog/Shutterstock, ANDREW LAMBERT PHOTOGRAPHY/SCIENCE PHOTO LIBRARY; p44 SCIENCE PHOTO LIBRARY; p45 SCIENCE PHOTO LIBRARY; p49 HUANSHENG XU/Shutterstock, Andraž Cerar/Shutterstock;

Chapter 2
p54 Chaiwuth Wichitdho/Shuttestock, Andraž Cerar/Shutterstock, bouybin/Shutterstock, Arsenis Spyros/Shutterstock; p55 yongyut rukkachatsuwa/Shutterstock, fivespots/Shutterstock, Shaiith/Shutterstock, izzzy71/ Shutterstock; p56 Shi Yali/Shutterstock, Lori Werhane/Shutterstock, izzzy71/ Shutterstock; p59 YuriyK/Shutterstock; p66 ANDREW LAMBERT PHOTOGRAPHY/SCIENCE PHOTO LIBRARY; p73 Jens Ottoson/Shutterstock; p74 iceink/Shutterstock; p76 Tatiana Popova/Shutterstock, science photo/Shutterstock; p77 Timur Djafarov/Shutterstock, Flegere/ Shutterstock; p78 Yeko Photo Studio/Shutterstock; p79 Georgi Roshkov/Shutterstock; p80 Vladimir A Veljanovski/Shutterstock; p81 Aumm graphixphoto/Shutterstock; p82 AlexanderAlUS/Shutterstock, nobeastsofierce/Shutterstock; p84 Syda Productions/Shutterstock;

Chapter 3
p92 llsshaya/Shutterstock, eldar nurkovic/Shutterstock, EM Karuna/Shutterstock, yurok/Shutterstock; p93 thodonal88/Shutterstock; p98 SCIENCE PHOTO LIBRARY; p99 GIPhotoStock/SCIENCE PHOTO LIBRARY, p100 Gjermund/Shutterstock; p104 Ming-Hsiang Chuang/ Shutterstock; p106 li jianbing/Shutterstock; p108 Robert Kneschke/Shutterstock;

Chapter 4
p118 jordache/Shutterstock, Dino Osmic/Shutterstock, Cales de Llierca, Sabine Kappel/Shutterstock, Slavoljub Pantelic/Shutterstock; p119 jordache/Shutterstock, Charles. D. Winters/SCIENCE PHOTO LIBRARY; p120 Kaband/Shutterstock; p121 JERRY MASON/SCIENCE PHOTO LIBRARY; P122 TREVOR CLIFFORD PHOTOGRAPHY/SCIENCE PHOTO LIBRARY, Filip Fuxa/Shutterstock; p124 Jiri Vaclavek/Shutterstock, Nastya22/Shutterstock; p125 jordache/Shutterstock; p126 GIPhotoStock/SCIENCE PHOTO LIBRARY; p130 NagyDodo/Shutterstock; p136 Dionisvera/Shutterstock, Ivaschenko Roman/Shutterstock, Slavoljub Pantelic/Shutterstock, igor.stevanovic/Shutterstock, Zern Liew/ Shutterstock; p144 CHOKCHAI POOMICHAIYA/Shutterstock;

Chapter 5
p158 gualtiero boffi/Shutterstock, Dino Osmic/Shutterstock, Dziewul/Shutterstock; p173 Albert Russ/Shutterstock; p160 Andrey_Popov/ Shutterstock; p166 Albert Russ/Shutterstock, Eugene Sergeev/Shutterstock;

Chapter 6
p174 Charles. D. Winters/SCIENCE PHOTO LIBRARY, Petrova Maria/Shutterstock, jmarkow/shutterstock, ggw1962/Shutterstock; p177 MARTYN F. CHILLMAID/SCIENCE PHOTO LIBRARY; p180 iladm/Shutterstock, RUI FERREIRA/Shutterstock; p192 ANDREW LAMBERT PHOTOGRAPHY/SCIENCE PHOTO LIBRARY, MARTYN F. CHILLMAID/SCIENCE PHOTO LIBRARY; p196 Dennis Sabo/Shutterstock;

Chapter 7
p208 Ulga/Shutterstock, PHOTO FUN/Shutterstock, stockphoto mania/Shutterstock, iurii/Shutterstock; p209 Charles Knowles/Shutterstock, A. Aleksandravicius/Shutterstock, Olivier Le Queinec/Shutterstock, ANDREW LAMBERT PHOTOGRAPHY/SCIENCE PHOTO LIBRARY; p210 iurii/Shutterstock; p214 Ulga/Shutterstock; p216 ggw1962/Shutterstock, PHOTO FUN/Shutterstock; p218 pixinoo/Shutterstock; p219 Olivier Le Queinec/Shutterstock, ANDREW LAMBERT PHOTOGRAPHY/SCIENCE PHOTO LIBRARY;

Chapter 8
p228 Swapan Photography/Shutterstock, ggw1962/shutterstock, STILLFX/Shutterstock, MARTYN F. CHILLMAID/SCIENCE PHOTO LIBRARY anyaivanova/Shutterstock; p229 ggw1962/Shutterstock, Italianvideophotoagency/Shutterstock, piximage/Shutterstock; ANDREW LAMBERT PHOTOGRAPHY/SCIENCE PHOTO LIBRARY p230 Lakeview Images/Shutterstock; p231 ANDREW LAMBERT PHOTOGRAPHY/SCIENCE PHOTO LIBRARY; p232 STILLFX/Shutterstock;

Chapter 9
p246 Ammit Jack/Shutterstock, Gualberto Becerra/Shutterstock, Jan Martin Will/Shutterstock, Hung Chung Chih/Shutterstock, Laurence Gough/Shutterstock; p247 Khoroshunova Olga/Shutterstock, Unicus/Shutterstock, Filip Fuxa/Shutterstock; Looker_Studio/Shutterstock; p250 Ammit Jack/Shutterstock, James Steidl/Shutterstock; p251 Khoroshunova Olga/Shutterstock; p252 Unicus/Shutterstock, attem/ Shutterstock; p253 Ekaterina Pokrovsky/Shutterstock; p254 Lee Prince/Shutterstock, M Rutherford/Shutterstock, schankz/Shutterstock; p259 Dudarev Mikhail/Shutterstock, Huguette Roe/Shutterstock; p260 Sohel Parvez Haque/Shutterstock, Silken Photography/Shutterstock, Saikat Paul/Shutterstock; p261 Matty Symons/Shutterstock; p262 Filip Fuxa/Shutterstock, martin33/Shutterstock, Tomasz Darul/ Shutterstock; p266 donikz/Shutterstock; p267 Warren Price Photography/Shutterstock; p268 Alexander Raths/Shutterstock, MikeDotta/ Shutterstock; p269 Petr Vopenka/Shutterstock, J. Helgason/Shutterstock, Laurence Gough/Shutterstock;

Chapter 10
p276 sculpies/Shutterstock, Artisticco/Shutterstock, Matee Nuserm/Shutterstock; p277 science photo/Shutterstock, JuliusKielaitis/ Shutterstock, Alessandro Colle/Shutterstock; p278 freedom100m/Shutterstock; p279 Olga Danylenko/Shutterstock, spwidoff/Shutterstock, p280 Designua/Shutterstock, ventdusud/Shutterstock, muratart/Shutterstock; p281 goodcat/Shutterstock; p284 Reddogs/Shutterstock; p285 Vaclav Volrab/shutterstock; p286 Mark Schwettmann/Shutterstock, Zoonar RF/ThinkStock MAXIMILIAN STOCK LTD/SCIENCE PHOTO LIBRARY; p287 Jose Arcos Aguilar/Shutterstock; p289 wavebreakmedia/Shutterstock, xshot/Shutterstock; p290 Pavel L Photo and Video/ Shutterstock, freedomnaruk/Shutterstock; p291 Therina Groenewald/Shuttestock, Huguette Roe/Shutterstock;

Contents

How to use this book

Learning objectives which are Higher tier only appear in a purple background box.

These tell you what you will be learning about in the lesson and are linked to the AQA specification.

This introduces the topic and puts the science into an interesting context.

Each topic is divided into three sections. The level of challenge gets harder with each section.

Remember! to cover all the content of the AQA Chemistry Specification you should study the text and attempt the End of Chapter Questions.

Chemistry

Electrolysis of molten ionic compounds

Learning objectives:

- identify which ions migrate to the cathode and anode
- explain how the ions of a molten electrolyte are discharged
- predict the products of electrolysis of molten binary compounds.

KEY WORDS

anode
cathode
discharged
molten

We have already learned that ions need to be free to move in an electrolyte if electrolysis is to take place. Ions that are not free to move cannot conduct electricity. Discharging the elements from compounds of highly reactive metals can happen if the substances are melted first.

Simple binary electrolytes

A simple binary electrolyte is one that is made up of two ions, for example, lead bromide or copper chloride.

These simple binary electrolytes can conduct electricity if melted. If they are not melted no electricity will flow.

This can be seen by setting up this apparatus in a laboratory, in a fume cupboard.

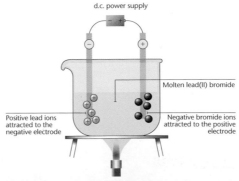

Figure 4.45 Electrolysis of molten lead bromide in the lab

When the lead bromide is cold and solid no electricity will flow and the lamp will not light.

If the crucible is heated and the solid begins to melt a current will flow and the lamp lights.

4.12

The ions, because they are now free to move, will migrate towards the electrodes.

1 Lead makes a positive ion. To which electrode will it move?

Positive and negative electrodes

When **molten** lead bromide is electrolysed using inert electrodes the ions will migrate towards them.

The lead ion, Pb^{2+}, is positive so will migrate towards the negative electrode.

The negative electrode is the cathode so a lead ion is a cation.

The ion is **discharged** and the metal lead, Pb, is produced at the **cathode**.

The non-metal bromide ion, Br^- is negative so will migrate towards the positive electrode.

The positive electrode is the anode so a bromide ion is an anion.

Two ions are discharged and the non-metal gas bromine, Br_2, is produced at the **anode**.

Another molten binary electrode that can be used in the lab is molten copper chloride.

The ions produced are Cu^{2+} and Cl^-.

2 To which electrode will a chloride ion move?

3 Predict the products of the electrolysis of molten sodium bromide at each electrode.

KEY INFORMATION

The electrodes used are inert. That means they do not react with any element that is discharged during electrolysis. They are normally made of carbon/graphite.

DID YOU KNOW?

Bromine is a brown gas that is a toxic irritant that causes burns. It has an unpleasant smell and bleaching action. That is why you should use a fume cupboard if it is produced.

HIGHER TIER ONLY

Electrode half equations

The ions move towards the electrodes and transfer electrons.

At the cathode	At the anode
Electrons are gained	Electrons are given up

For $PbBr_2$ the half equations are:

$$Pb^{2+} + 2e^- \rightarrow Pb \qquad 2Br^- - 2e^- \rightarrow Br_2$$

Other decompositions can be written as half equations if the formulae of the ions are known.

Electrolyte	Half equation at cathode	Half equation at anode
KCl	$2K^+ + 2e^- \rightarrow K$	$2Cl^- - 2e^- \rightarrow Cl_2$
$CuCl_2$	$Cu^{2+} + 2e^- \rightarrow Cu$	$2Cl^- - 2e^- \rightarrow Cl_2$
PbI_2	$Pb^{2+} + 2e^- \rightarrow Pb$	$2I^- - 2e^- \rightarrow I_2$
Al_2O_3	$2Al^{3+} + 6e^- \rightarrow 2Al$	$6O^{2-} - 12e^- \rightarrow 3O_2$

4 Write the half equations at the cathode and anode for the electrolysis of molten copper bromide.

Google search: 'electrolysis of molten lead bromide'   143

Chemistry

The first page of a chapter has links to ideas you have met before, which you can now build on.

This page gives a summary of the exciting new ideas you will be learning about in the chapter.

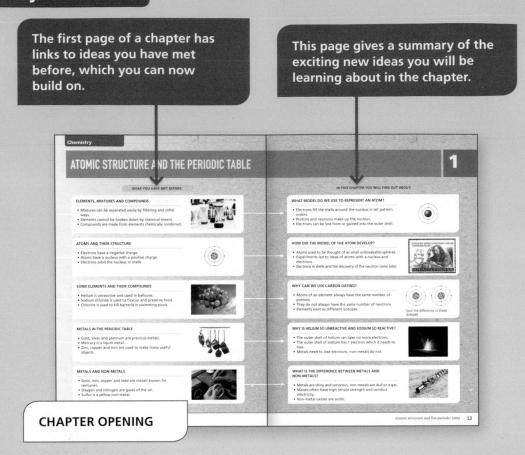

CHAPTER OPENING

The Key Concept pages focus on a core ideas. Once you have understood the key concept in a chapter, it should develop your understanding of the whole topic.

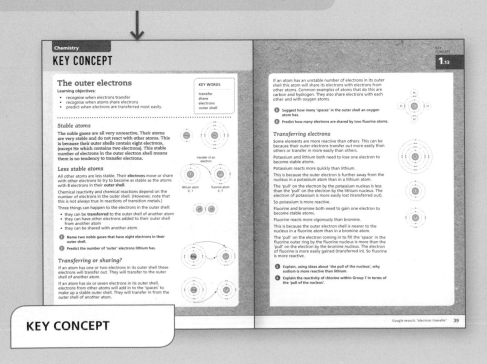

KEY CONCEPT

There is a dedicated page for every Required Practical in the AQA specification. They help you to analyse the practical and to answer questions about it.

The tasks – which get a bit more difficult as you go through – challenge you to apply your science skills and knowledge to the new context.

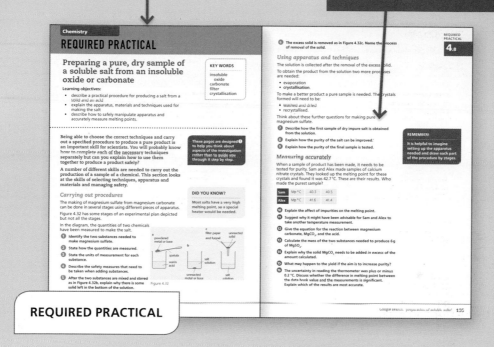

REQUIRED PRACTICAL

The Maths Skills pages focus on the maths requirements in the AQA specification, explaining concepts and providing opportunities to practise.

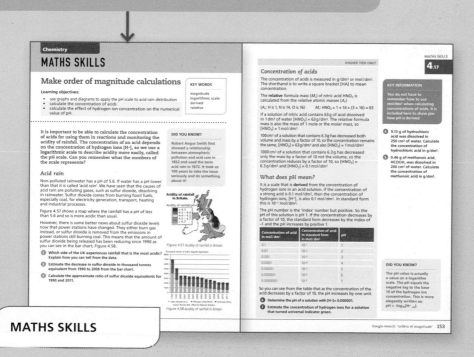

MATHS SKILLS

Chemistry

These lists at the end of a chapter act as a checklist of the key ideas of the chapter. In each row, the green box gives the ideas or skills that you should master first. Then you can aim to master the ideas and skills in the blue box. Once you have achieved those you can move on to those in the red box.

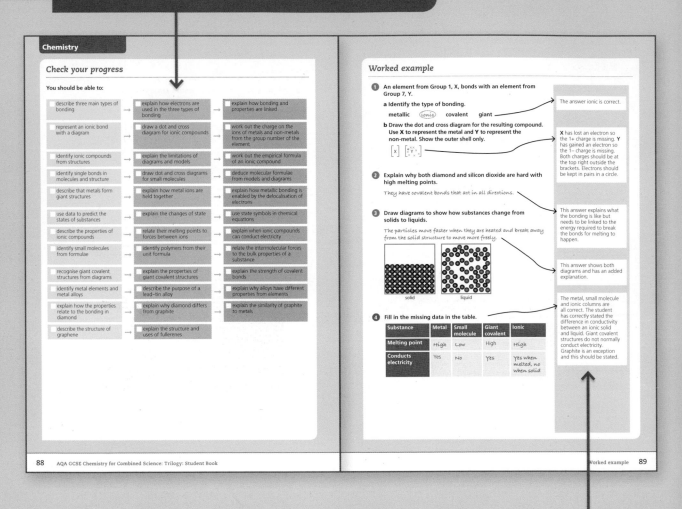

Check your progress

You should be able to:

describe three main types of bonding	explain how electrons are used in the three types of bonding	explain how bonding and properties are linked
represent an ionic bond with a diagram	draw a dot and cross diagram for ionic compounds	work out the charge on the ions of metals and non-metals from the group number of the element
identify ionic compounds from structures	explain the limitations of diagrams and models	work out the empirical formula of an ionic compound
identify single bonds in molecules and structure	draw dot and cross diagrams for small molecules	deduce molecular formulae from models and diagrams
describe that metals form giant structures	explain how metal ions are held together	explain how metallic bonding is enabled by the delocalisation of electrons
use data to predict the states of substances	explain the changes of state	use state symbols in chemical equations
describe the properties of ionic compounds	relate their melting points to forces between ions	explain when ionic compounds can conduct electricity
identify small molecules from formulae	identify polymers from their unit formula	relate the intermolecular forces to the bulk properties of a substance
recognise giant covalent structures from diagrams	explain the properties of giant covalent structures	explain the strength of covalent bonds
identify metal elements and metal alloys	describe the purpose of a lead–tin alloy	explain why alloys have different properties from elements
explain how the properties relate to the bonding in diamond	explain why diamond differs from graphite	explain the similarity of graphite to metals
describe the structure of graphene	explain the structure and uses of fullerenes	

Worked example

1 An element from Group 1, X, bonds with an element from Group 7, Y.

a Identify the type of bonding.

metallic   (ionic)   covalent   giant

> The answer ionic is correct.

b Draw the dot and cross diagram for the resulting compound. Use **X** to represent the metal and **Y** to represent the non-metal. Show the outer shell only.

> **X** has lost an electron so the 1+ charge is missing. **Y** has gained an electron so the 1– charge is missing. Both charges should be at the top right outside the brackets. Electrons should be kept in pairs in a circle.

2 Explain why both diamond and silicon dioxide are hard with high melting points.

They have covalent bonds that act in all directions.

> This answer explains what the bonding is like but needs to be linked to the energy required to break the bonds for melting to happen.

3 Draw diagrams to show how substances change from solids to liquids.

The particles move faster when they are heated and break away from the solid structure to move more freely.

solid   liquid

> This answer shows both diagrams and has an added explanation.

4 Fill in the missing data in the table.

Substance	Metal	Small molecule	Giant covalent	Ionic
Melting point	High	Low	High	High
Conducts electricity	Yes	No	Yes	Yes when melted, no when solid

> The metal, small molecule and ionic columns are all correct. The student has correctly stated the difference in conductivity between an ionic solid and liquid. Giant covalent structures do not normally conduct electricity. Graphite is an exception and this should be stated.

Use the comments to help you understand how to answer questions. Read each question and answer. Try to decide if, and how, the answer can be improved. Finally, read the comments and try to answer the questions yourself.

END OF CHAPTER

The End of Chapter Questions allow you and your teacher to check that you have understood the ideas in the chapter, can apply these to new situations, and can explain new science using the skills and knowledge you have gained. The questions start off easier and get harder. If you are taking Foundation tier try to answer all the questions in the Getting started and Going further sections. If you are taking Higher Tier try to answer all the questions in the Going further, More challenging and Most demanding sections.

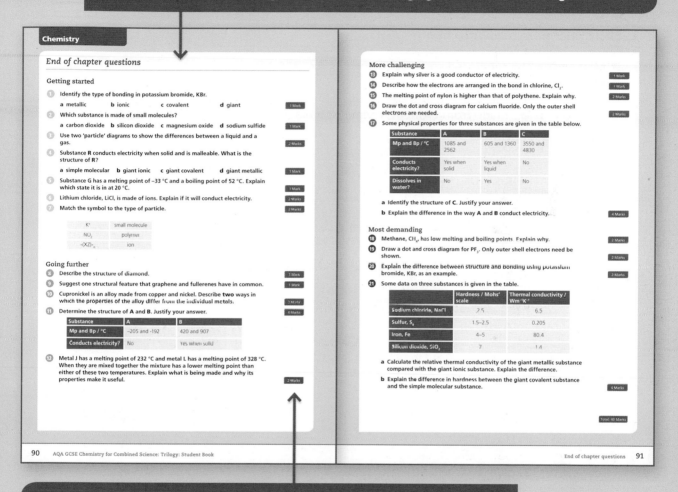

End of chapter questions

Getting started

1 Identify the type of bonding in potassium bromide, KBr.
   a metallic   b ionic   c covalent   d giant   **1 Mark**

2 Which substance is made of small molecules?
   a carbon dioxide b silicon dioxide c magnesium oxide d sodium sulfide   **1 Mark**

3 Use two 'particle' diagrams to show the differences between a liquid and a gas.   **2 Marks**

4 Substance R conducts electricity when solid and is malleable. What is the structure of R?
   a simple molecular b giant ionic c giant covalent d giant metallic   **1 Mark**

5 Substance G has a melting point of –33 °C and a boiling point of 52 °C. Explain which state it is in at 20 °C.   **1 Mark**

6 Lithium chloride, LiCl, is made of ions. Explain if it will conduct electricity.   **2 Marks**

7 Match the symbol to the type of particle.   **2 Marks**

K⁺	small molecule
NO₂⁻	polymer
-(XZ)-ₙ	ion

Going further

8 Describe the structure of diamond.   **1 Mark**

9 Suggest one structural feature that graphene and fullerenes have in common.   **1 Mark**

10 Cupronickel is an alloy made from copper and nickel. Describe **two** ways in which the properties of the alloy differ from the individual metals.   **2 Marks**

11 Determine the structure of **A** and **B**. Justify your answer.   **4 Marks**

Substance	A	B
Mp and Bp / °C	–205 and –192	420 and 907
Conducts electricity?	No	Yes when solid

12 Metal J has a melting point of 232 °C and metal L has a melting point of 328 °C. When they are mixed together the mixture has a lower melting point than either of these two temperatures. Explain what is being made and why its properties make it useful.   **2 Marks**

More challenging

13 Explain why silver is a good conductor of electricity.   **1 Mark**

14 Describe how the electrons are arranged in the bond in chlorine, Cl_2.   **1 Mark**

15 The melting point of nylon is higher than that of polythene. Explain why.   **2 Marks**

16 Draw the dot and cross diagram for calcium fluoride. Only the outer shell electrons are needed.   **2 Marks**

17 Some physical properties for three substances are given in the table below.

Substance	A	B	C
Mp and Bp / °C	1085 and 2562	605 and 1360	3550 and 4830
Conducts electricity?	Yes when solid	Yes when liquid	No
Dissolves in water?	No	Yes	No

   a Identify the structure of **C**. Justify your answer.
   b Explain the difference in the way **A** and **B** conduct electricity.   **4 Marks**

Most demanding

18 Methane, CH_4, has low melting and boiling points. Explain why.   **2 Marks**

19 Draw a dot and cross diagram for PF_3. Only outer shell electrons need be shown.   **2 Marks**

20 Explain the difference between structure and bonding using potassium bromide, KBr, as an example.   **2 Marks**

21 Some data on three substances is given in the table.

	Hardness / Mohs' scale	Thermal conductivity / Wm⁻¹K⁻¹
Sodium chloride, NaCl	2.5	6.5
Sulfur, S₈	1.5–2.5	0.205
Iron, Fe	4–5	80.4
Silicon dioxide, SiO₂	7	1.4

   a Calculate the relative thermal conductivity of the giant metallic substance compared with the giant ionic substance. Explain the difference.
   b Explain the difference in hardness between the giant covalent substance and the simple molecular substance.   **4 Marks**

Total: 40 Marks

There are questions for each assessment objective (AO) from the final exams. These help you to develop the thinking skills you need to answer each type of question.

AO1 – to answer these questions you should aim to **demonstrate** your knowledge and understanding of scientific ideas, techniques and procedures.

AO2 – to answer these questions you should aim to **apply** your knowledge and understanding of scientific ideas and scientific enquiry, techniques and procedures.

AO3 – to answer these questions you should aim to **analyse** information and ideas to: interpret and evaluate, make judgements and draw conclusions, develop and improve experimental procedures.

ATOMIC STRUCTURE AND THE PERIODIC TABLE

IDEAS YOU HAVE MET BEFORE:

ELEMENTS, MIXTURES AND COMPOUNDS

- Mixtures can be separated easily by filtering and other ways.
- Elements cannot be broken down by chemical means.
- Compounds are made from elements chemically combined.

ATOMS AND THEIR STRUCTURE

- Electrons have a negative charge.
- Atoms have a nucleus with a positive charge.
- Electrons orbit the nucleus in shells.

SOME ELEMENTS AND THEIR COMPOUNDS

- Helium is unreactive and used in balloons.
- Sodium chloride is used to flavour and preserve food.
- Chlorine is used to kill bacteria in swimming pools.

METALS IN THE PERIODIC TABLE

- Gold, silver and platinum are precious metals.
- Mercury is a liquid metal.
- Zinc, copper and iron are used to make many useful objects.

METALS AND NON-METALS

- Gold, iron, copper and lead are metals known for centuries.
- Oxygen and nitrogen are gases of the air.
- Sulfur is a yellow non-metal.

IN THIS CHAPTER YOU WILL FIND OUT ABOUT:

WHAT MODEL DO WE USE TO REPRESENT AN ATOM?

- Electrons fill the shells around the nucleus in set pattern orders.
- Protons and neutrons make up the nucleus.
- Electrons can be lost from or gained into the outer shell.

HOW DID THE MODEL OF THE ATOM DEVELOP?

- Atoms used to be thought of as small unbreakable spheres.
- Experiments led to ideas of atoms with a nucleus and electrons.
- Electrons in shells and the discovery of the neutron came later.

WHY CAN WE USE CARBON DATING?

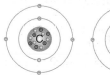

- Atoms of an element always have the same number of protons.
- They do not always have the same number of neutrons.
- Elements exist as different isotopes.

Spot the difference in these *isotopes*

WHY IS HELIUM SO UNREACTIVE AND SODIUM SO REACTIVE?

- The outer shell of helium can take no more electrons.
- The outer shell of sodium has 1 electron which it needs to lose.
- Metals need to lose electrons, non-metals do not.

WHAT IS THE DIFFERENCE BETWEEN METALS AND NON-METALS?

- Metals are shiny and sonorous, non-metals are dull or a gas.
- Metals often have high tensile strength and conduct electricity.
- Non-metal oxides are acidic.

Elements and compounds

Learning objectives:

- identify symbols of elements from the periodic table
- recognise compounds from their formula
- identify the elements in a compound.

KEY WORDS

....

compound
element

All the elements are listed in the periodic table. The elements in the formula of any compound, no matter how large, can be identified by using the periodic table.

Elements and compounds

An **element** is a substance that cannot be broken down chemically.

A **compound** is a substance that contains at least two different elements, chemically combined in fixed proportions.

		PERIODIC TABLE					
$^1_1$ H hydrogen		ELEMENTS 1-20					$^4_2$ He helium
$^7_3$ Li lithium	$^9_4$ Be beryllium	$^{11}_5$ B boron	$^{12}_6$ C carbon	$^{14}_7$ N nitrogen	$^{16}_8$ O oxygen	$^{19}_9$ F fluorine	$^{20}_{10}$ Ne neon
$^{23}_{11}$ Na sodium	$^{24}_{12}$ Mg magnesium	$^{27}_{13}$ Al aluminium	$^{28}_{14}$ Si silicon	$^{31}_{15}$ P phosphorus	$^{32}_{16}$ S sulfur	$^{35}_{17}$ Cl chlorine	$^{40}_{18}$ Ar argon
$^{39}_{19}$ K potassium	$^{40}_{20}$ Ca calcium						

Figure 1.1 Two sections of the periodic table. Can you find magnesium and oxygen?

Figure 1.2 Magnesium (metal) reacts with oxygen (gas) to make magnesium oxide (compound).

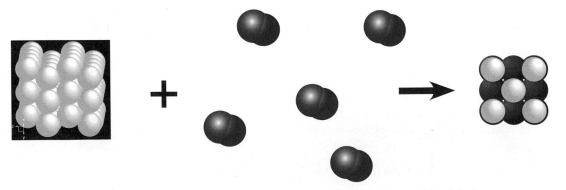

Figure 1.3 The reaction of the elements in Figure 1.2 can be represented by models of their atoms.

1 **Identify the following substances as elements or compounds.**

  copper   copper chloride   copper sulfate

2 **Name the elements in beryllium chloride.**

Compounds and elements

Copper is an element. It cannot be broken down into any other substances, but salt can.

The chemical name for common salt is sodium chloride. Sodium chloride is a compound. Sodium chloride can be broken down to make sodium and chlorine, but this is not easy to do because the sodium and chlorine are chemically combined. We need to use electricity to make sodium and chlorine from sodium chloride.

Sodium and chlorine cannot be broken down any further. Sodium and chlorine are elements.

There are only about 100 elements but these can join together chemically to make an enormous number of compounds. They need chemical reactions to do this.

KEY INFORMATION

When chlor<u>ine</u> reacts to make a compound it chemically combines and becomes a chlor<u>ide</u>.

Similarly, bromine reacts to become a bromide and oxygen reacts to become an oxide.

3 Identify the elements in potassium bromide.

4 Predict the products when lead iodide is split by electricity.

Making the element copper into a compound

A chemical reaction is needed to make copper (an element) into a compound. If copper is burned in oxygen it forms the compound copper oxide. Chemical reactions always involve the formation of one or more new substances, and often involve a detectable energy change.

Figure 1.4 Copper (an element) burning in oxygen (an element) to make copper oxide (a compound). This is normally done by heating the copper in crucibles.

DID YOU KNOW?

Compounds can only be separated into elements by chemical reactions. To get the element copper back, the oxygen needs to be chemically removed. This is done using hydrogen. The oxygen combines with the hydrogen to make water, another compound.
copper oxide + hydrogen → copper + water

5 Name the compound made from sodium and oxygen.

6 Oxygen can be removed from iron(III) oxide by carbon monoxide. Identify the element and compound produced.

7 Substance D reacted with hydrogen to form zinc and water. Explain whether substance D is an element or a compound.

Atoms, formulae and equations

KEY WORDS

balanced
compound
element
equation
formula
molecule
symbol

Learning objectives:

- explain that an element consists of the same type of atoms
- explain that atoms join together to make molecules
- explain how formulae represent elements and compounds.

All substances are chemicals. Many people say "I don't want chemicals in my food" not realising that foods *are* chemicals. Our food is made of compounds and mixtures. Compounds are elements joined together in many different ways. In fact, we are all made of chemical compounds.

Atoms and molecules

Elements are made up of atoms that are all the same.

Compounds are made up of atoms (or charged atoms) of different elements which have been chemically joined together.

CH_3COOH is a compound made from three elements. These are carbon, C, hydrogen, H, and oxygen, O. You will know it as vinegar. Its chemical name is ethanoic acid. The atoms join by sharing their electrons.

Figure 1.5 Vinegar is used in salad dressing. Its chemical name is ethanoic acid. Ethanoic acid is a compound made from carbon, hydrogen and oxygen.

ethanoic acid

Figure 1.7 The molecule in vinegar, CH_3COOH

Figure 1.6 Gold is an element. Carbon monoxide is a compound. How could you tell by looking at these diagrams?

If two or more atoms join together by sharing their electrons the atoms form a **molecule**.

Two examples of molecules are oxygen and water.

oxygen molecule   water molecule

Figure 1.8 Which of these is a molecule of an element? How can you tell?

1 Explain whether the following substances are elements or compounds.

C (carbon), CO_2 (carbon dioxide), Cl_2 (chlorine) and SO_3 (sulfur trioxide)

2 For each of the substances listed below, give:

a the names of the elements

b how many different types of atoms they contain

c how many atoms are in the molecule overall.

CO_2 S_8 Cl_2 SO_3 C_{60} C_4H_{10}

KEY INFORMATION

In a chemical formula the number that is a subscript on the bottom right is the number of atoms in the molecule, for example C_2H_6 has two carbon atoms joined to six hydrogen atoms.

Formulae

You can see which elements are in a compound by looking at its **formula**. For example, the compound magnesium oxide (MgO) contains Mg (magnesium) and O_2 (oxygen).

Look back at the section of the periodic table on page 14.

Elements from Group 1 and elements from Group 7 combine to make compounds in a fixed ratio of 1 : 1.

- The formula of lithium fluoride is LiF.

Elements from Group 2 and elements from Group 6 also combine to make compounds in a fixed ratio of 1 : 1.

- The formula of calcium oxide is CaO.

Elements from Group 2 and elements from Group 7 combine to make compounds in a fixed ratio of 1 : 2. The element that you need two of has a suffix '2' after the symbol.

- The formula of calcium fluoride is CaF_2.

Elements from Group 1 and elements from Group 6 combine to make compounds in a fixed ratio of 2 : 1. Again, the element that you need two of has a suffix '2' after the symbol.

- The formula of sodium sulfide is Na_2S.

3 Give the names of the elements in $MgSO_4$ and in CH_4.

4 Determine the formulae of lithium chloride, magnesium chloride and potassium oxide.

Equations and balancing

When magnesium reacts with oxygen it makes magnesium oxide.

The word **equation** is:

magnesium + oxygen → magnesium oxide

The **symbol** equation is:

Mg	+	O_2	→	MgO	
1		2		1	1 ✗

This is not a **balanced** symbol equation because there are two atoms of oxygen in the reactants and only one atom of oxygen in the products. We need to add '2' to the front of the formulae of magnesium and magnesium oxide. This gives:

2Mg	+	O_2	→	2MgO	
2		2		2	2 ✓

5 Write a balanced equation for the formation of sodium chloride from sodium, Na, and chlorine, Cl_2.

6 Write a balanced equation for the formation of aluminium oxide, Al_2O_3.

7 Complete and balance the following equation by suggesting values for **D**, **E** and **F**:

$$\textbf{D } N_2 + \textbf{E } O_2 \rightarrow \textbf{F } NO_2$$

Mixtures

Learning objectives:

- recognise that all substances are chemicals
- understand that all substances are either mixtures, compounds or elements
- explain that mixtures can be separated.

KEY WORDS

chromatography
filtration
mixture
separation

You will have begun to use a range of equipment to safely separate chemical mixtures, and we need to extend this range of techniques. Filtering and distillation are probably familiar but fractional distillation can also be used to separate mixtures.

Mixtures

Many substances are made of mixtures. **Mixtures** can easily be separated because the chemicals in them are not joined together. Mixtures can be separated by **filtration**, crystallisation, simple distillation, fractional distillation and **chromatography**.

Figure 1.9 Separating mixtures

Let's take a mixture of salt and copper. To separate them we add water to the mixture. The salt dissolves but the copper does not. The salt solution can be filtered through a filter paper, leaving the copper behind as a residue. The salt solution can be crystallised to make solid salt crystals. These physical processes do not involve chemical reactions and no new substances are made.

After separating, salt and copper can be mixed again.

1. **Draw a diagram of the equipment used to filter salt solution from copper.**

2. **Explain why a blend of copper and salt is not a compound.**

Separating mixtures

A mixture consists of two or more elements or compounds not chemically combined together. The chemical properties of each substance in the mixture are unchanged. Mixtures can be separated by physical processes. These processes do not involve chemical reactions.

These **separation** processes include:

- filtration
- crystallisation
- distillation
- chromatography

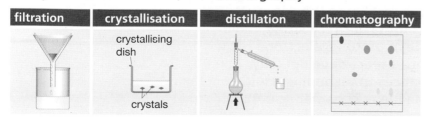

Figure 1.10 Techniques for separating mixtures

KEY INFORMATION

Filtration separates insoluble substances from soluble substances and distillation separates liquids that have different boiling points.

3 Which technique would be used to separate coloured inks in a mixture?

4 Which technique would be used to separate alcohol (boiling point 80 °C) and water?

Fractional distillation

Some mixtures are very complex and have many different components. These mixtures can be separated either (a) by using different techniques in sequence or (b) by the same technique which includes multiple separations.

An example of the first approach is filtration followed by crystallisation.

An example of the second approach is fractional distillation, where different liquids with different boiling points are separated at different points in the process.

Fractional distillation works by using a tall tower of gaps and surfaces, which are gradually colder towards the top. The liquid mixture is heated at the bottom and the liquids boil together to make a mixture of gases. As each gas reaches a surface at the same temperature as its boiling point (or condensing point), that gas will condense and the condensed liquid will run off. The other gases continue up through the gaps until they reach the surface at their condensing temperature. Eventually nearly all the gases in the mixture will condense and be collected as separated liquids. The final gas is left at the top of the tower and is collected as a gas.

5 Suggest how you would collect a specimen of clean copper sulfate crystals from a mixture of solid copper sulfate, sand and alcohol.

6 Suggest the order (from bottom to top) in which these liquids would be collected from the fractional distillation tower:

- **a** (boiling point 85 °C)
- **c** (boiling point 35 °C)
- **b** (boiling point 100 °C)
- **d** (boiling point 165 °C)

DID YOU KNOW?

Fractional distillation is used to separate the different substances in crude oil.

fractional distillation

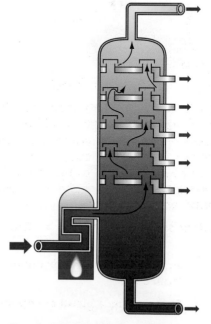

Figure 1.11 Fractional distillation of mixtures with different boiling points

Changing ideas about atoms

Learning objectives:

- describe how the atomic model has changed over time
- explain why the atomic model has changed over time
- understand that a theory is provisional until the next piece of evidence is available.

KEY WORDS

Ernest Rutherford
Geiger and Marsden
J. J. Thomson
James Chadwick
John Dalton
Niels Bohr

The idea of atoms has changed hugely over the years. At the moment, scientists believe atoms are very small, have a very small mass and are made of protons, electrons and neutrons. Our current theories were developed by imagination, evidence and advances in technology, with each new idea being built on the ideas of earlier scientists.

Developing the atomic theory

Explanations about atoms began about 400 BC, when the Greek philosopher Democritus described materials as being made of small particles which could not be divided. He called these particles 'atoms'. However, he had no evidence. It was just an idea.

Little more was suggested for more than 2000 years, but in 1803 the British scientist **John Dalton** used his observations to describe the atom in more detail. His model described an atom as a 'billiard ball' which could not be divided.

Figure 1.12 Dalton's idea of atoms: they were like tiny billiard balls.

Dalton's model was then changed as new evidence was found.

In 1897, 94 years later, **J. J. Thomson** discovered the electron. Thomson developed the way that the atom was thought of by using a 'plum pudding' model to describe atoms. Negative electrons were thought to be embedded in a ball of positive charge, rather like the fruit (the electrons) are part of a pudding (the ball of positive charge).

KEY INFORMATION

At each stage, the explanations of atomic theory were provisional until more convincing evidence was found to make the model better.

1. Suggest why Dalton's atomic model did not include positive and negative charge.

2. Explain why the discovery of the electron changed the Dalton model of the atom.

Changing theories

Sometimes ideas can develop rapidly because of unexpected results.

In 1909 **Geiger** and **Marsden** had really surprising results in their experiment with gold leaf and alpha particles. These results led Geiger, Marsden and **Ernest Rutherford** to propose a new idea that an atom has a nucleus. In 1911, Rutherford suggested the atom had a positively charged nucleus in the centre where most of the mass of the atom was concentrated and much of the atom was empty space. This was the nuclear model of the atom.

Figure 1.13 Rutherford and Geiger in their lab in Manchester, UK.

In 1913, **Niels Bohr** used theoretical calculations that agreed with experimental evidence to adapt the nuclear model. He explained that the electrons orbited the nucleus in definite orbits at specific distances from the nucleus. He explained that a fixed amount of energy (a *quantum* of energy) is needed for an electron to move from one orbit to the next. Electrons only exist in these orbits.

3 Suggest why Bohr proposed that electrons orbited the nucleus in shells.

4 What is meant by the phrase 'quantum of energy'?

Further development of atomic theory

Later experiments gradually led to the idea that the positive charge of any nucleus can be sub-divided into a whole number of smaller particles. Each of these particles had the same amount of positive charge. In 1920 the term 'proton' was first used in print for these particles.

In 1932, **James Chadwick** discovered the neutron. Again this discovery involved experimental evidence and mathematical analysis.

5 Draw a timeline of the discoveries that led to our present understanding of the atomic theory.

6 Suggest why it was twelve years between finding protons and neutrons.

DID YOU KNOW?

As a challenge you can find out about the Geiger and Marsden experiment that changed the theory from a 'plum-pudding' atom to a nuclear atom, it is a famous turning point in the understanding of atoms.

DID YOU KNOW?

The idea of atoms as small particles is not new. However, our ideas about the theory of atoms are still developing. Search on 'CERN LHC' to find out more.

Modelling the atom

Learning objectives:

- describe the atom as a positively charged nucleus surrounded by negatively charged electrons
- explain that most of the mass of an atom is in the nucleus
- explain that the nuclear radius is much smaller than that of the atom and most of the mass is in the nucleus.

KEY WORDS

charge
electron
electron shell
negative
nucleus
positive

Atoms are the building blocks of all matter, both living and non-living, simple and complex. Atoms join together in millions of different ways to make *all* the materials around us. We can explain how *everything*, including ourselves, is made by using ideas and models of atoms.

Atoms

Individual atoms are very small. There are about ten million million atoms in this full stop.

An atom is made up of a **nucleus** that is surrounded by **electrons**.

- The nucleus carries the **positive charge**.
- Electrons, which surround the nucleus, each carry a **negative** charge.

It depends how it is measured, but the diameter of an atom is about 10^{-10} m. That's 0.000 000 1 mm. If we imagine that an atom is blown up to the size of a football stadium the nucleus would be the size of a peanut placed on the centre spot.

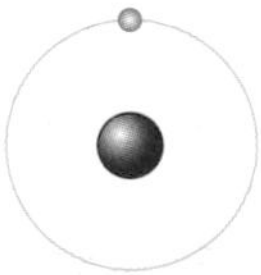

1 What is the type of charge in the nucleus?

2 Helium has two positive charges in the nucleus. Predict the number of electrons in a helium atom.

More on atoms

Electrons occupy the space around the nucleus in 'shells'. The space between the nucleus and the **electron shells** is empty space.

Figure 1.14 Image of gold atoms. Magnification x 16 000 000.

Figure 1.15 The structure of a hydrogen atom. What charge does an electron carry?

The nucleus contains most of the mass of the atom and the electrons contribute very little. On the other hand, the radius of the atom, where the electrons are orbiting, is much larger than the radius of the nucleus in the centre.

When we are talking about these differences we are talking about small sizes. Atoms are *very* small. A typical atomic radius is about 0.1 nm (1×10^{-10} m). The radius of a nucleus is less than one ten-thousandth of the radius of an atom (about 1×10^{-14} m).

Typical atomic radius	Typical radius of a nucleus
1×10^{-10} m	1×10^{-14} m

The radius of an atom is measured in many different ways. This is because the outer electron shell is not a fixed boundary, and so its position can only be measured approximately.

3 Most of the atom is empty space. What does this suggest about the size of an electron?

4 Explain why the radius of the nucleus is much smaller than the radius of the whole atom.

DID YOU KNOW?

An atom of gold has a mass of about 3.3×10^{-22} g and a radius of about 1.4×10^{-10} m. Most of the mass of the atom is in the middle, in the nucleus.

HIGHER TIER ONLY

The radius of an atom increases within a group of elements.

For example the atomic radii of Li, Na, K, increase as more electrons are 'added' to the atom.

5 Suggest why the radius of potassium is larger than the radius of lithium.

6 The positive charge on a Li nucleus is 3. The positive charge on a Ne nucleus is 10. As more negative electrons are added one by one to atoms from Li up to Ne the radius gets smaller, not bigger. Suggest why. Use ideas about opposite charges.

KEY INFORMATION

Remember that the typical radius of a nucleus is less than 1/10 000 th of the typical radius of an atom.

Relating charges and masses

Learning objectives:

* describe the structure of atoms
* recall the relative masses and charges of protons, neutrons and electrons
* explain why atoms are neutral.

We have seen how ideas about atoms have changed over the years. Currently, scientists believe atoms are made of three important particles – protons, electrons and neutrons.

* The numbers of **protons** and **neutrons** are important in *nuclear reactions*.
* The number of **electrons** is important in *chemical reactions*.

Structure of atoms

An atom is made up of a nucleus that is surrounded by electrons.

* The nucleus carries a positive charge.
* The electrons that surround the nucleus each carry a negative charge.

The nucleus of an atom is made up of protons and neutrons.

* Protons have a positive charge.
* Neutrons have no charge.

An atom always has the same number of protons (+) as electrons (–) so atoms are always **neutral**.

The **atomic number** is the number of protons in an atom. The atomic number for helium is 2 because it has two protons.

1. Lithium has an atomic number of 3. Predict the number of electrons in lithium.

2. The neon atom has 10 protons. Explain why the neon atom is neutral.

3. Use the periodic table to identify the element with 3 protons.

4. Determine the number of protons in an atom of calcium, Ca.

Masses and charges

The nucleus of an atom is made up of particles (protons and neutrons) that are much heavier than electrons. The relative masses and charges of electrons, protons and neutrons are shown in the table.

DID YOU KNOW?

Even these protons and neutrons can be broken down further in huge particle accelerators such as the one built deep underneath Switzerland by a joint team of scientists and engineers from many European countries.

KEY INFORMATION

It is because a helium atom has two protons that it has an atomic number of 2.

Figure 1.16 The structure of a helium atom. There are the same number of protons and electrons.

	Relative charge	Relative mass
Electron	−1	0.0005
Proton	+1	1
Neutron	0	1

5 A fluorine atom has 9 positive charges, 9 negative charges and a mass of 19. Describe the structure of its atom.

6 A chlorine atom has 17 electrons and a mass of 35. Describe the structure of its atom.

Losing electrons

If an atom has an atomic number of 3 and a neutral charge, it must be a lithium atom. It has a neutral charge because the atom has three protons (+) and three electrons (−).

If the lithium atom loses one negatively charged electron it then becomes a *charged* particle with one positive charge that is not balanced out by a negative charge.

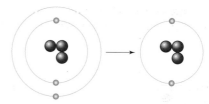

Figure 1.17 A neutral lithium atom loses an electron and becomes charged.

	Atomic number	Number of protons	Number of electrons	Charge
Lithium atom	3	3	3	0
Lithium charged particle	3	3	2	ı1

If an atom loses electrons and becomes charged, this charged particle is called a positive *ion*.

7 If a magnesium atom loses 2 electrons it becomes a charged particle. It still has a mass of 24. Write out the atomic number, number of protons, number of electrons, number of neutrons and the charge of

a a magnesium atom

b a magnesium ion.

8 Explain why a magnesium atom is neutral but a magnesium ion is charged.

9 Nitride ions have a −3 charge. Work out the number of electrons in a nitride ion, given that the atomic number of nitrogen is 7.

Sub-atomic particles

Learning objectives:

- use the definition of atomic number and mass number
- calculate the numbers of protons, neutrons and electrons in atoms
- calculate the numbers of sub-atomic particles in isotopes and ions.

KEY WORDS

relative atomic mass
isotope
neutrons
protons

Smoke detectors, archaeological dating and bone imaging all use isotopes. Some elements have more than one type of atom. These different types of atom have different numbers of neutrons and are called isotopes.

Atomic number and mass number

The nucleus of an atom is made up of **protons** and **neutrons**.

- The atomic number is the number of protons in an atom.
- The *mass number* of an atom is the total number of protons and neutrons in an atom.

If a particle has an atomic number of 11, a mass number of 23 and a neutral charge, it must have:

- 11 protons, because it has an atomic number of 11
- 11 electrons, because there are 11 protons and the atom is neutral
- 12 neutrons, because the mass number is 23 and there are already 11 protons (23 – 11 = 12).

Here are some more examples.

	Atomic number	Mass number	Number of protons	Number of electrons	Number of neutrons
Carbon	6	12	6	6	6
Fluorine	9	19	9	9	10
Sodium	11	23	11	11	12
Aluminium					

1 Complete the row for an atom of aluminium, Al.

2 Work out the numbers of protons, electrons and neutrons in an atom with an atomic number of 15 and a mass number of 31.

Isotopes

All atoms of carbon have 6 protons, so its atomic number is 6. Most carbon atoms have 6 neutrons, so the mass number is 6 + 6 = 12. This form of the carbon atom is written as $^{12}_{6}C$.

Another form of carbon, $^{14}_{6}C$, has an atomic number of 6 (6 protons) and a mass number of 14. It must therefore have 8 neutrons (14 – 6). $^{14}_{6}C$ is sometimes written as carbon-14. $^{12}_{6}C$ and $^{14}_{6}C$ are **isotopes** of carbon.

3 Write the isotope symbol for an atom that has 17 protons and 18 neutrons.

4 Identify all the sub-atomic particles in an atom of carbon-13.

Relative abundance of isotopes

Most elements have two or more isotopes. For example, hydrogen has three common isotopes.

Isotope	Electrons	Protons	Neutrons	Mass number
$^{1}_{1}H$	1	1	0	1
$^{2}_{1}H$	1	1	1	2
$^{3}_{1}H$	1	1	2	3

The **relative atomic mass** of an element is the average mass of the different isotopes of an element. Chlorine's A_r of 35.5 is an average of the masses of the different isotopes of chlorine.

There are two main isotopes, $^{35}_{17}Cl$ and $^{37}_{17}Cl$. If there were 50% of each of the isotopes what would be the average mass? The answer is 36. But there is less of the $^{35}_{17}Cl$ isotope. So we need a *relative abundance* calculation:

$$A_r = \frac{\left(\begin{array}{c}\text{mass of first isotope}\\ \times\% \text{ of first isotope}\end{array}\right) + \left(\begin{array}{c}\text{mass of second isotope}\\ \times\% \text{ of second isotope}\end{array}\right)}{100}$$

For example, for chlorine the abundance values are:

75% $^{35}_{17}Cl$ and 25% $^{37}_{17}Cl$

Therefore:

$$A_r = \frac{(75 \times 35) + (25 \times 37)}{100}$$
$$= \frac{2625 + 925}{100}$$
$$= 35.5$$

5 Explain the similarities and differences between the three isotopes of hydrogen.

6 Element X has two isotopes, mass 27 and 29. Calculate the relative atomic mass of X if the first isotope has an abundance of 65% and the second isotope has 35% abundance.

Electronic structure

Learning objectives:

* explain how electrons occupy 'shells' in order
* describe the pattern of the electrons in shells for the first 20 elements.

KEY WORDS

electronic
  structure
electron shells
energy levels
group

The electrons of an atom are arranged in patterns. The electrons fill up shells in order until that shell can take no more electrons. The next electron goes into the next shell. These patterns are the key to the behaviour of atoms.

The 'build-up' of electrons

Electrons occupy shells around the nucleus. The **electron shell** nearest to the nucleus takes up to two electrons. The second shell takes up to eight electrons. The next electrons occupy a third shell.

Oxygen has an atomic number of 8. It has eight protons and so it has eight electrons in the space around the nucleus. The first two go into the first shell. As the first shell is now full, the next 6 electrons go into the second shell. The electron pattern for oxygen is then 2,6.

1 Draw the electron pattern for hydrogen and for lithium.

2 Write down the electron pattern for nitrogen.

The shells are not fixed rings and are also known as **energy levels**.

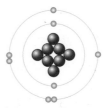

Figure 1.18 For oxygen, the eight electrons can be written as 2,6 or drawn like this.

Electron patterns and groups

The periodic table is arranged in order of 'proton number'.

There is a very important link between **electronic structure** and the periodic table. For example, let us consider an atom of an element that has the electronic structure 2,8,6.

* This element has three electron shells, so it is in the third row of the periodic table.
* It has six electrons in its outer shell, so it is in the sixth column.
* Using the periodic table, we can find that its atomic number is 16 and it is the element sulfur, S.

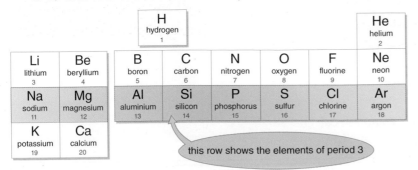

Figure 1.19 Period 3 contains the elements from sodium to argon.

We can also work the other way. Find the element with the atomic number 12 in the periodic table. This is magnesium, Mg.

- It is in the third row, so it has three electron shells.
- It is in the second column, so it has two electrons in its outer shell.
- It has the electronic structure 2,8,2.

The column of elements is known as a **group**. So column 2 is Group 2.

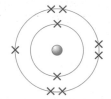

Figure 1.20 An atom of fluorine, 2,7

3 **An atom of an element has an atomic number of 11.**

   a **Draw a diagram to show the pattern of electrons.**

   b **Identify the element.**

   c **Identify the group to which it belongs.**

4 **Work out the electronic structure of the element that has an atomic number of 9.**

5 **An element has a mass number of 40 and an electron arrangement of 2,8,8,2. Identify the element and work out the number of neutrons.**

Maximum numbers

The electronic structure of each of the first 20 elements can be worked out using:

- the atomic number of the element
- the maximum number of electrons in each shell.

The third shell takes up to eight electrons before the fourth shell starts to fill. Element 19, potassium, has the electronic structure 2,8,8,1.

Use the periodic table to help you to answer these questions.

6 **Work out the electronic structure of argon.**

7 **Use a blank periodic table sheet to draw out the electronic structure of the first 20 elements, putting the diagram in the correct box. What do you notice about the group number and the number of electrons in the outer shell?**

8 **The electronic configuration for the ion of an unknown isotope $^{26}X^{3+}$ is 2,8.**

   a **Work out the atomic number of element X.**

   b **Determine the number of neutrons in X.**

   c **Explain which group in the periodic table element X is in.**

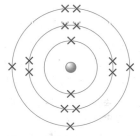

Figure 1.21 An atom of phosphorus, 2,8,5

DID YOU KNOW?

It took many years for scientists to work out the theory of electrons occupying shells. They started with the behaviour of elements and then looked for patterns.

KEY INFORMATION

Do not try to use this method to work out the electronic structure of gold, as it has 79 electrons.

The periodic table

KEY WORDS

electron shells
energy levels
group
period

Learning objectives:

- explain how the electronic structure of atoms follows a pattern
- recognise that the number of electrons in an element's outer shell corresponds to the element's group number
- explain that the electronic structure of transition metals positions the elements into the transition metal block.

We know that the periodic table is arranged in rows and columns and the elements are written in order of their atomic number. So why are all the elements in the last column unreactive gases? Why are all the elements in the first column highly reactive metals? The answer lies in the pattern of their electrons.

The order of elements and electron patterns

As we have seen, the elements in the periodic table are arranged in order of atomic number. Atomic number is the number of protons in an atom. As atoms are neutral, the atomic number also gives the number of electrons in an atom.

For example, the atomic number of hydrogen is 1, carbon is 6 and sodium is 11. This means that hydrogen is the first element in the table, carbon is the sixth and sodium is the eleventh.

We have also seen that electrons occupy **energy levels** (or shells). Each element has a pattern of electrons (known as its electronic structure) that is built up in a particular order.

The electronic structure of each of the first 20 elements can be worked out using:

- the atomic number of the element
- the maximum number of electrons allowed in each shell.

The third shell takes up to eight electrons before the fourth shell starts to fill. Element 20, therefore, has the electronic structure 2,8,8,2.

1 Which element has the atomic number 13?

2 Which element has an electronic structure of 2,8,7?

Arrangement of groups and periods

A row of the periodic table is known as a **period**. The first row of the periodic table contains the elements hydrogen and helium. This row is called the first period. These two elements only have electrons in the first shell (energy level).

Lithium's third electron goes into the next electron shell. Lithium starts a new row in the periodic table. This second row is called the second **period**.

KEY INFORMATION

Remember:

- the first shell of electrons carries up to 2 electrons
- the second shell carries up to 8 electrons.
- the third shell carries up to 8 electrons and the fourth up to 18.

Figure 1.22 The electron pattern of a lithium atom

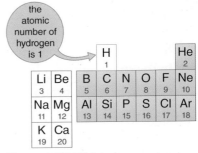

Figure 1.23 Lithium has an atomic number of 3. It has three protons and three electrons.

Let's consider an atom of the element that has the electronic structure 2,8,2. We can work out that it is in the third row of the periodic table. This atom has three **electron shells**. It has two electrons in its outer shell. Its atomic number is 12. (You can work this out by adding up the numbers of electrons.) Looking at the periodic table, the atom is of the element magnesium, 12.

So the element is therefore, in the third **period**.

Looking again, this atom only has two electrons in its outer shell, so it is in the second column. A column is known as a **group**. This second column is known as Group 2.

The group number refers to the number of electrons in the outside shell.

We have seen that the electron pattern for Li is 2,1. We can work out that the pattern for Na (11 electrons) is 2,8,1 and K (19 electrons) is 2,8,8,1. Both of these elements have one electron in their outside shell. They are both in the first column. This column is known as Group 1.

3 What is the pattern of electrons in the atom of the element with atomic number 16? Identify the element, its group and its period.

4 To which group and period does the element chlorine belong?

Electronic structure and behaviour of elements

Element 20 has the electronic structure 2,8,8,2 and is in Group 2.

The fourth shell can take up to 18 electrons, but the next element (element number 21) is not in Group 3. Instead the next 10 elements (numbers 21 to 30) in period 4 are part of the first *transition element period*. Transition elements have characteristics that are similar to each other. For example, even though they are all metals, iron behaves more like copper than sodium.

Elements in groups often behave similarly to other elements in that group:

- the elements in Group 1 are all highly reactive metals
- the elements in Group 7 (the halogens) all react with Group 1 metals to make salts
- the elements in Group 0 are all unreactive (or noble) gases.

This pattern in reactivity occurs because the *elements in each group have the same number of electrons in their outer shells*.

Use the periodic table to help you to answer these questions.

5 What are the electronic structures of F and Cl? Why are they both found in Group 7?

6 Identify the elements in Group 6.

1.9

KEY INFORMATION

The periodic table is arranged in rows (called periods). Each period number is the same as the number of electron shells. Each period contains the elements whose outside shell of electrons is 'filling up'.

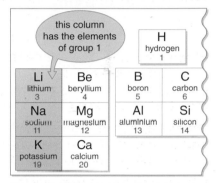

Figure 1.24 These elements are in the first column as they all have one electron in their outside shell.

Developing the periodic table

Learning objectives:

- describe the steps in the development of the periodic table
- explain how Mendeleev left spaces for undiscovered elements
- explain why the element order in the modern periodic table was changed
- explain how testing a prediction can support or refute a new scientific idea.

Although some elements have been known about as parts of compounds since ancient times, many were only first isolated as elements in the last two centuries. For a long time, scientists trying to find patterns were working with an incomplete picture. It was like trying to do a 100-piece jigsaw puzzle with half of the pieces missing.

51 23 V	52 24 Cr	?	56 26 Fe	59 27 Co	59 28 Ni	?	65 30 Zn
vanadium	chromium		iron	cobalt	nickel		zinc
?	96 42 Mo	99 43 Tc	101 44 Ru	?	106 46 Pd	108 47 Ag	112 48 Cd
	molybdenum	technetium	ruthenium		palladium	silver	cadmium
181 73 Ta	184 74 W	?	190 76 Os	192 77 Ir	?	197 79 Au	?
tantalum	tungsten		osmium	iridium		gold	

Figure 1.25 The element puzzle

The timeline of ideas

Some elements have been known since ancient times.

By 1829 Döbereiner had noticed that sometimes three elements had similar **properties**. He noticed **patterns** with:

lithium, sodium and potassium	calcium, strontium and barium	chlorine, bromine and iodine

Figure 1.26 Döbereiner

These were called 'Döbereiner triads'.

In 1860 a new list of more accurate atomic weights was published.

In 1865 John Newlands noticed that when he put the elements in atomic weight order (even though some then seemed to be in the wrong place) there was often a pattern of similar properties every eight elements. He called his new theory the 'law of octaves'.

Döbereiner and Newlands noticed *patterns* in the properties of elements but did not make **predictions**.

1 Describe the 'law of octaves'.

2 Suggest why there were so few elements for Döbereiner to test his theory on in 1829.

Allowing for predictions

By 1869, the Russian scientist Dmitri Mendeleev had also put the elements in order of their atomic weights. He also saw that some elements seemed to be in the wrong order. But crucially he:

- decided to swap some elements round so that the patterns of chemical behaviour fitted better
- was able to imagine that there were undiscovered elements
- brilliantly decided to leave gaps in his **periodic table** for later discoveries and used the patterns of chemical behaviour to decide where to leave these gaps.

He gave special names to these unknown elements in his gaps. He took the name of the element above the gap in that group and put the prefix 'eka' in front of the name. One unknown was eka-silicon (beyond silicon). This element was eventually discovered and was named germanium. Altogether three of these new elements were discovered within Mendeleev's lifetime.

Figure 1.27 Dmitri Mendeleev

3 Mendeleev named one element eka-aluminium. It was later discovered. Identify the element.

4 Explain why some elements appeared to be in the wrong order.

Discovering the unpredictable

Mendeleev died in 1907 and so did not live to see the discovery of sub-atomic particles, which gave the final evidence allowing the modern periodic table to be developed in the order of the *atomic number* of elements.

His theory was developed in 1869, and it was finally supported with evidence 63 years later, in 1932. The evidence of isotopes finally explained why the order based on atomic weight was incorrect. The discovery of the neutron explained why the order of elements in the modern periodic table needs to be by the number of protons.

5 Mendeleev left gaps and made predictions for undiscovered elements in the his periodic table. Explain how the discovery of germanium supported this approach.

6 Search for an image of a new periodic table.

  a Give the atomic number of the last element before the central block begins.

  b Identify the missing elements in Figure 1.25.

7 Identify two elements in the periodic table that would be in the wrong position if they were ordered by atomic mass.

8 A student stated that 'The modern periodic table is complete'. Briefly explain whether you agree with the student.

> **REMEMBER!**
>
> You should aim to be able to explain that if germanium had not fitted into Mendeleev's pattern then the evidence would have *refuted* his predictions, but instead, it did fit his pattern so the new evidence *supported* his predictions.

> **DID YOU KNOW?**
>
> Mendeleev did not win a Nobel Prize; however, he does have an element named after him. Look up which number this is.

Comparing metals and non-metals

Learning objectives:

- recall a number of physical properties of metals and non-metals
- describe some chemical properties of metals and non-metals
- explain the differences between metals and non-metals on the basis of their characteristic physical and chemical properties.

KEY WORDS

electrical
  conductivity
lustrous
tensile strength
thermal
  conductivity

Metals are useful materials as they have a wide range of properties. Gold does not corrode and it has an appealing colour and lustre, so it is used for jewellery. Copper has good thermal conductivity so is used for saucepans. Non-metal elements are mostly combined in compounds to be useful.

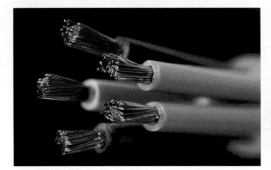  

Figure 1.28 Many metals are instantly recognisable as metals because they are shiny and sonorous.

Physical properties

Most metals are instantly recognisable as they are shiny whereas non-metals are not. These are the other physical properties of metals and non-metals.

Metals	Non-metals
lustrous	dull
hard (with the exception of mercury which is a liquid at room temperature)	soft, brittle, liquids or gas (for most non-metals at room temperature)
high density	low density
high **tensile strength**	low or no tensile strength or gas
high melting point and boiling point	low melting point and boiling point
good conductors of heat	poor or no **thermal conductivity**
good **electrical conductivity**	poor or non-conductors of electricity (with the exception of carbon)

Figure 1.29 Dating from 1779, this is the first bridge made from cast iron. Iron was used to make this bridge because it is very strong.

Figure 1.30 Sulfur and nitrogen are non-metals. They are not shiny and have low boiling points.

1 Write down two physical properties of silver that make it more useful than sulfur for making cutlery.

2 Use the picture in Figure 1.30 and the table to explain why sulfur is not a metal.

Chemical properties

Chemical properties of metals are the result of reactions with oxygen or acids. Although copper is resistant to attack by oxygen and acid (which is a reason why it is used for saucepans), many other metals react with oxygen to make an oxide (such as calcium oxide and iron oxide).

Metals react with acids to make salts and hydrogen gas.

One chemical property of a non-metal is the result of the reaction with oxygen. Carbon reacts with oxygen to make carbon dioxide. Carbon dioxide dissolves in water to make a mildly acidic solution. This is what fizzy water contains.

3 Suggest a test to show that a solution of carbon dioxide is mildly acidic.

4 Predict the products of magnesium and nitric acid.

Figure 1.31 Bottles of lemonade and cola with acidic carbon dioxide solution

Distinguishing properties

Sulfur and phosphorus both react with oxygen to make oxides. Both sulfur dioxide and phosphorus oxide turn universal indicator red. They are acidic oxides.	Calcium and potassium both react with oxygen to make oxides. Both calcium oxide and potassium oxide turn universal indicator blue. They are basic oxides.

Non-metals form acidic (or neutral) oxides. Metals form basic oxides.

5 You have a sample of 'unknownium' oxide. Explain how you would use universal indicator to see if 'unknownium' was a metal or a non-metal.

6 An element makes an oxide that turns universal indicator red. Is the element a metal or a non-metal?

7 Element X has a low melting point of 63 °C, has low density, is a good electrical conductor and is malleable. Explain whether the oxide of X is likely to be acidic or basic.

DID YOU KNOW?

Other physical properties of metals include being malleable and ductile.

KEY INFORMATION

You will soon need to know how to explain the chemical properties of metals and non-metals using knowledge about how they form ions.

1.11

Metals and non-metals

Learning objectives:

- describe that metals are found on the left of the periodic table and non-metals on the right
- explain the differences between metals and non-metals based on their physical and chemical properties
- explain that metals form positive ions and non-metals do not.

KEY WORDS

ions
atomic structure
metalloids

Whether an element is a metal or a non-metal depends on the electronic structure of its atoms. The element is classified as a non-metal or a metal depending on whether it needs to gain or lose electrons and if some of its reactions are typical for a non-metal or a metal.

Metals and non-metals in the periodic table

Li lithium 3				N nitrogen 7	O oxygen 8		Ne neon 10
Na sodium 11	Mg magnesium 12			P phosphorus 15	S sulfur 16	Cl chlorine 17	
	Ca calcium 20						

Figure 1.32 Here are some metals and non-metals in the periodic table.

Looking at the periodic table you can see the metals lithium, sodium, magnesium and calcium on the left-hand side.

You can see the non-metals nitrogen, oxygen, sulfur, phosphorus, chlorine and neon on the right-hand side.

From your knowledge of chemistry so far, try to draw a line that separates the metals from the non-metals.

DID YOU KNOW?

The line of elements separating the metals from the non-metals are called the metalloids.

Group 1	Group 2											Group 3	4	5	6	7	0
			$^{1}_{1}$H hydrogen														$^{4}_{2}$He helium
$^{7}_{3}$Li lithium	$^{9}_{4}$Be beryllium											$^{11}_{5}$B boron	$^{12}_{6}$C carbon	$^{14}_{7}$N nitrogen	$^{16}_{8}$O oxygen	$^{19}_{9}$F fluorine	$^{20}_{10}$Ne neon
$^{23}_{11}$Na sodium	$^{24}_{12}$Mg magnesium											$^{27}_{13}$Al aluminium	$^{28}_{14}$Si silicon	$^{31}_{15}$P phosphorus	$^{32}_{16}$S sulfur	$^{35}_{17}$Cl chlorine	$^{40}_{18}$Ar argon
$^{39}_{19}$K potassium	$^{40}_{20}$Ca calcium	$^{45}_{21}$Sc scandium	$^{48}_{22}$Ti titanium	$^{51}_{23}$V vanadium	$^{52}_{24}$Cr chromium	$^{55}_{25}$Mn manganese	$^{56}_{26}$Fe iron	$^{59}_{27}$Co cobalt	$^{59}_{28}$Ni nickel	$^{64}_{29}$Cu copper	$^{65}_{30}$Zn zinc	$^{70}_{31}$Ga gallium	$^{73}_{32}$Ge germanium	$^{75}_{33}$As arsenic	$^{79}_{34}$Se selenium	$^{80}_{35}$Br bromine	$^{84}_{36}$Kr krypton
$^{85}_{37}$Rb rubidium	$^{88}_{38}$Sr strontium	$^{89}_{39}$Y yttrium	$^{91}_{40}$Zr zirconium	$^{93}_{41}$Nb niobium	$^{96}_{42}$Mo molybdenum	$^{99}_{43}$Tc technetium	$^{101}_{44}$Ru ruthenium	$^{103}_{45}$Rh rhodium	$^{106}_{46}$Pd palladium	$^{108}_{47}$Ag silver	$^{112}_{48}$Cd cadmium	$^{115}_{49}$In indium	$^{119}_{50}$Sn tin	$^{122}_{51}$Sb antimony	$^{128}_{52}$Te tellurium	$^{127}_{53}$I iodine	$^{131}_{54}$Xe xenon

Figure 1.33 Top section of the periodic table

1 **Find element 53. Is this element a metal or a non-metal?**

2 **Is element 26 a metal or a non-metal?**

Positions of elements in the table

Magnesium is a metal because of its **atomic structure**. It has two electrons in the outer shell, which can be easily lost from the atom. The electrons join another atom that needs more electrons in its outer shell. Because electrons have been lost, a positive **ion** has been made.

Magnesium makes a positive ion so it is a metal.

3 **Explain why aluminium is a metal, using knowledge about its atomic structure.**

4 **Explain why element number 8 is a non-metal. Use ideas about atomic structure.**

Electron transfer in metals and non-metals

When a metal, such as magnesium, reacts with a non-metal such as oxygen, the metal loses electrons and the non-metal gains electrons.

magnesium   +   oxygen   →   magnesium oxide

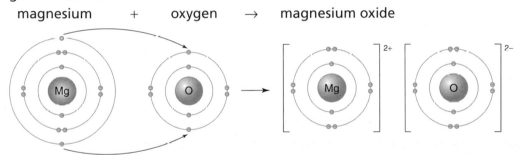

Figure 1.35

This is because metal atoms have a few outer electrons which they 'lose' to form positive ions. Oxygen accepts the electrons.

When a non-metal, such as chlorine, reacts with a metal such as sodium, the non-metal gains an electron from the metal.

sodium   +   chlorine   →   sodium chloride

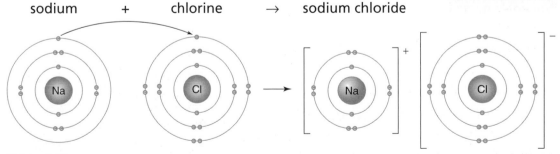

Figure 1.36

This is because non-metal atoms have empty spaces in their outer shell in which they 'gain' other electrons from metals to form negative ions. The non-metal does not form a positive ion.

5 **Explain why fluorine is a non-metal that can react with the metal, potassium, to form potassium fluoride.**

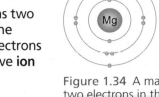

Figure 1.34 A magnesium atom with two electrons in the outer shell

DID YOU KNOW?

Metal atoms are bonded together by their outer electrons. The atoms pack together and the outer electrons delocalise, which means that the outer electrons move through the ions as a 'sea' of electrons.

KEY INFORMATION

Metals make positive ions and non-metals do not.

6 **The elements with atomic number 3 and 9 can react together. Explain why and work out the formula of the product.**

KEY CONCEPT

The outer electrons

Learning objectives:

- recognise when electrons transfer
- recognise when atoms share electrons
- predict when electrons are transferred most easily.

KEY WORDS
....
transfer
share
electrons
outer shell

Stable atoms

The noble gases are all very unreactive. Their atoms are very stable and do not normally react with other atoms. This is because their outer shells contain eight electrons, (except He which contains two electrons). This stable number of electrons in the outer electron shell means there is no tendency to transfer electrons.

Less stable atoms

All other atoms are less stable. Their **electrons** move or **share** with other electrons to try to become as stable as the atoms with 8 electrons in their **outer shell**.

Chemical reactivity and chemical reactions depend on the number of electrons in the outer shell. (However, note that this is not always true in reactions of transition metals.)

Three things can happen to the electrons in the outer shell:

- they can be **transferred** to the outer shell of another atom
- they can have other electrons added to their outer shell from another atom
- they can be shared with another atom.

1 **Name two noble gases that have eight electrons in their outer shell.**

2 **Predict the number of 'outer' electrons lithium has.**

Transferring or sharing?

If an atom has one or two electrons in its outer shell these electrons will transfer out. They will transfer to the outer shell of another atom.

If an atom has six or seven electrons in its outer shell, electrons from other atoms will add in to the 'spaces' to make up a stable outer shell. They will transfer in from the outer shell of another atom.

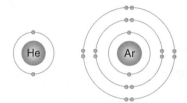

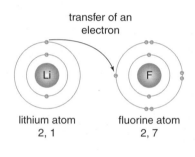

transfer of an electron

lithium atom
2, 1

fluorine atom
2, 7

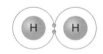

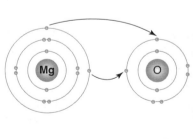

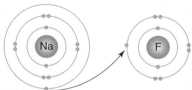

If an atom has an unstable number of electrons i.e. 3, 4 or 5 in its outer shell this atom will share its electrons with electrons from other atoms. Common examples of atoms that do this are carbon and hydrogen. They also share electrons with each other and with oxygen atoms.

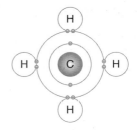

3 Suggest how many 'spaces' in the outer shell an oxygen atom has.

4 Predict how many electrons are shared by two fluorine atoms.

Transferring electrons

Some elements are more reactive than others. This can be because their outer electrons transfer *out* more easily than others or transfer *in* more easily than others.

Potassium and lithium both need to lose one electron to become stable atoms.

Potassium reacts more quickly than lithium.

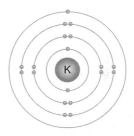

This is because the outer electron is further away from the nucleus in a potassium atom than in a lithium atom.

The 'pull' on the electron by the potassium nucleus is less than the 'pull' on the electron by the lithium nucleus. The electron of potassium is more easily lost (transferred out).

So potassium is more reactive.

Fluorine and bromine both need to gain one electron to become stable atoms.

Fluorine reacts more vigorously than bromine.

This is because the outer electron shell is nearer to the nucleus in a fluorine atom than in a bromine atom.

The 'pull' on the electron coming in to fill the 'space' in the fluorine outer ring by the fluorine nucleus is more than the 'pull' on the electron by the bromine nucleus. The electron of fluorine is more easily gained (transferred in). So fluorine is more reactive.

5 Explain, using ideas about 'the pull of the nucleus', why sodium is more reactive than lithium.

6 Explain the reactivity of chlorine within Group 7 in terms of the 'pull of the nucleus'.

Exploring Group 0

Learning objectives:

- describe the unreactivity of the noble gases
- predict and explain the trend in boiling point of the noble gases (going down the group)
- explain how properties of the elements in Group 0 depend on the outer shell of electrons of their atoms.

KEY WORDS
....
elements
helium
neon
argon
unreactive

Ever wondered why helium rather than hydrogen is used in party balloons and for weather balloons? Hydrogen is less dense than helium and so a hydrogen balloon would 'float' better. However, hydrogen is highly flammable whereas helium is unreactive. The decision comes down to safety.

Patterns in Group 0

If you find Group 0 on the periodic table, you will see these **elements** in order.

Group 0
Helium (He)
Neon (Ne)
Argon (Ar)
Krypton (Kr)
Xenon (Xe)

Figure 1.37 Helium balloons

All the elements of Group 0 in the periodic table have two things in common.

- They are all **unreactive**.
- They are all gases.

The boiling points of the elements in Group 0 show a trend.

Helium has the lowest boiling point. This means the atoms of the element keep moving rapidly (as a gas) at lower temperatures than the atoms of xenon.

Boiling points (°C)	
He	−268
Ne	−246
Ar	−186
Kr	−153
Xe	−108

The trend is that the boiling points of the gases increase down the group. Therefore, boiling points increase as atomic mass increases.

EXTENSION

....

Mendeleev did not predict the noble gases. They were discovered much later by William Ramsay. You need to be able to explain that if the noble gases had not fitted into Mendeleev's pattern then the evidence would have *refuted* his predictions but instead they did fit his pattern so the new evidence *supported* his predictions.

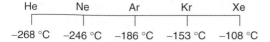

He   Ne   Ar   Kr   Xe

−268 °C −246 °C −186 °C −153 °C −108 °C

Figure 1.38

1. Find the relative atomic mass of neon and argon. Determine which has the higher relative atomic mass.

2. There is a noble gas with a higher relative atomic mass than Xe. Predict its boiling point.

Why does helium stay as a gas at lower temperatures?

All the elements in Group 0 are gases.

These gases exist as single atoms, not molecules.

The smaller the atom, the 'easier' it is for it to keep moving around rapidly.

So at lower temperatures the small atoms of helium move 'more easily' than the larger atoms of krypton. So krypton has a higher boiling point than helium.

3. Describe how boiling point varies with relative atomic mass for Group 0 elements.

4. Suggest a relationship between the diameter of Group 0 atoms and their boiling point. Explain your answer.

Why do elements in Group 0 exist as single atoms?

The elements of Group 0 do not make compounds with other elements and are unreactive.

They do not make compounds because the atoms have eight electrons in their outer shell.

This is a very stable configuration. So there is no electron movement from one atom to another.

5. Draw the electronic structure for helium. Explain why this atom does not join with any other atom.

6. Write out the electronic structure (in numbers) for argon. Explain why argon is unreactive but has a higher boiling point than helium.

7. One of the first noble gas compounds to be made was xenon tetrafluoride, made in 1962. It is a stable crystalline solid at room temperature.

   a Complete and balance the following equation:

   ___ Xe + ___ F_2 → _____

   b Suggest why it was a surprise that a noble gas compound had been made.

DID YOU KNOW?

Helium is also used with **neon** in the lasers that scan supermarket barcodes.

Street lights used in long tunnels contain both neon and sodium. **Argon** is used for specialist welding and to fill the space between double glazed windows.

Find out what krypton and xenon are used for.

Figure 1.39 Neon lights were first used years ago.

Figure 1.40 The electronic structure of neon. The outer shell is full.

Exploring Group 1

Learning objectives:

- explain why Group 1 metals are known as the alkali metals
- predict the properties of other Group 1 metals from trends down the group
- relate the properties of the alkali metals to the number of electrons in their outer shell.

KEY WORDS

alkali metals
dense
ion
reactivity
stable electronic
    structure

Ever wondered how fireworks are made to have such stunning colours? It is because they have specific compounds added to them. Sodium compounds, for instance, produce a bright yellow colour against the night sky. Sodium is an element in the periodic table in the first column.

Properties of Group 1 elements

Figure 1.41 Which metal gives the firework this lilac colour?

Figure 1.42 This is sodium on water. Why does it float?

Lithium, sodium and potassium are Group 1 elements that are less **dense** than water. All three metals would float on the surface when added to water.

Group 1 metals react vigorously with water and make hydrogen. Group 1 metals also burn in oxygen to form oxides. Sodium burns to make sodium oxide.

Figure 1.43 This is potassium when it burns in oxygen.

1. **Explain why sodium floats on water.**

2. **Identify the gas given off when potassium reacts with water.**

Reaction trends of alkali metals

When lithium, sodium and potassium react with water, hydrogen is given off. An alkaline solution of the metal hydroxide is also made in the reaction. Sodium forms sodium hydroxide.

Lithium reacts vigorously with water.

Sodium reacts very vigorously with water.

Potassium reacts extremely vigorously with water and produces a lilac flame.

The word equation for the reaction between sodium and water is:

sodium + water → sodium hydroxide + hydrogen

The balanced symbol equation for the reaction is:

$2Na + 2H_2O \rightarrow 2NaOH + H_2$

3 Suggest a way to show that the solution is an alkali.

4 Describe the trend in reactivity down Group 1.

5 Predict the reactivity of rubidium with water. Justify your answer.

Making ions

Alkali metals have similar chemical properties. This is because when they react their atoms need to lose one electron to form the electronic structure of a noble gas. This is then a **stable electronic structure**.

When the atom loses one electron it forms an **ion**. It has one more positive charge in its nucleus than negative electrons surrounding it. So it is now a positive ion that carries a charge of +1.

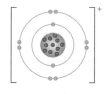

Figure 1.45 How alkali metals achieve a stable electronic structure

Sodium reacts with chlorine to make sodium chloride. The sodium makes an ion that carries a +1 charge. It makes an ionic compound. The compound is a white solid that dissolves in water to form a colourless solution. The other alkali metals make these compounds too.

6 Explain why Group 1 atoms lose electrons.

7 Draw a 'dot and cross' diagram to show an Li ion.

8 Potassium reacts with oxygen to form an oxide.

    a Write a balanced equation for the reaction.

    b State the electron configurations of potassium and oxygen in the oxide.

    c The oxide dissolves in water to form an alkaline solution. Identify this solution.

DID YOU KNOW?

Sodium hydroxide can be used as oven cleaner. This label warns us to avoid getting sodium hydroxide on our skin.

Figure 1.44 Why do the instructions on the bottle tell you to use gloves?

Sodium hydroxide is more damaging to the eyes than acids. The hydroxide ions travel to the back of the eyes and can irreversibly damage the retina. That is why your teacher will always tell you to wear safety glasses when handling chemicals, especially alkalis.

Exploring Group 7

Learning objectives:

- recall that fluorine, chlorine, bromine and iodine are non-metals called halogens
- describe that they react vigorously with alkali metals
- construct balanced symbol equations for the reactions of metals with halogens.

Group 7 elements are known as the halogens. The first use of chlorine was to bleach textiles and it is still used in chemicals for bleaching toilets. Chlorine is used to sterilise water, which prevents diseases such as cholera from spreading. Chlorine was also used as a weapon in the First World War. The effects were devastating.

The halogens

Group 7 elements, fluorine, **chlorine**, **bromine** and **iodine**, are called the **halogens**. The halogens all have 7 electrons in their outer shell and so all share similar chemical properties.

Halogens are non-metals. Halogens exist as pairs of atoms, so their symbols are F_2, Cl_2, Br_2 and I_2.

They react vigorously with **metals** such as sodium, potassium and magnesium to produce salts. For example:

Potassium reacts with chlorine to make potassium chloride. The word equation is:

potassium + chlorine → potassium chloride

It is possible to construct a balanced symbol equation for the reaction

$$K + Cl_2 \rightarrow KCl \text{ (unbalanced)}$$

$$2K + Cl_2 \rightarrow 2KCl \text{ (balanced)}$$

The halogens react with non-metals to make gases or liquids such as acids.

Group					0
					$^{4}_{2}$He helium
3	4	5	6	7	
$^{11}_{5}$B boron	$^{12}_{6}$C carbon	$^{14}_{7}$N nitrogen	$^{16}_{8}$O oxygen	$^{19}_{9}$F fluorine	$^{20}_{10}$Ne neon
$^{27}_{13}$Al aluminium	$^{28}_{14}$Si silicon	$^{31}_{15}$P phosphorus	$^{32}_{16}$S sulfur	$^{35}_{17}$Cl chlorine	$^{40}_{18}$Ar argon
$^{70}_{31}$Ga gallium	$^{73}_{32}$Ge germanium	$^{75}_{33}$As arsenic	$^{79}_{34}$Se selenium	$^{80}_{35}$Br bromine	$^{84}_{36}$Kr krypton
$^{115}_{49}$In indium	$^{119}_{50}$Sn tin	$^{122}_{51}$Sb antimony	$^{128}_{52}$Te tellurium	$^{127}_{53}$I iodine	$^{131}_{54}$Xe xenon
$^{204}_{81}$Tl thallium	$^{207}_{82}$Pb lead	$^{209}_{83}$Bi bismuth	$^{210}_{84}$Po polonium	$^{210}_{85}$At astatine	$^{222}_{86}$Rn radon

Figure 1.46 Group 7 elements in the periodic table

1. Give the balanced symbol equation for the reaction between sodium and bromine.

2. Identify the product of the reaction between lithium and fluorine.

Figure 1.47 Chlorine reacting with potassium to make potassium chloride

Group 7 trends

There is a **trend** in the physical appearance of the halogens at room temperature. Chlorine is a gas and iodine is a solid.

Figure 1.48 At room temperature chlorine is a green gas and iodine is a grey solid. What colour is the toxic and volatile bromine?

Halogen	Atomic mass	Relative molecular mass	Melting point (°C)	Boiling point (°C)	State at room temperature
Chlorine		71	−101	−34	gas
Bromine	80	160	−7	59	
Iodine	127		114	184	

3 Complete the data table.

4 Explain how the appearance of iodine in Figure 1.48 is confirmed by the Melting point column of the table.

5 Describe the trend in boiling point as molecular mass changes.

Displacement reactions of halogens

The **reactivity** of the halogens decreases down the group.

If halogens are bubbled through solutions of metal halides there are two possibilities: no reaction, or a **displacement reaction**.

If chlorine is bubbled through potassium bromide solution, a displacement reaction occurs. An orange colour of bromine is seen. This is because chlorine is more reactive than bromine.

6 Identify which halogens will displace iodine from a solution, of potassium iodide. Justify your answer.

7 Write a balanced symbol equation for the reaction between chlorine and potassium iodide.

8 Astatine is at the bottom of Group 7. It is radioactive and extremely rare so its chemistry has not been well studied.

a Predict the state of astatine at room temperature.

b Write a balanced equation for the reaction between chlorine and sodium astatide, NaAt.

c Explain whether astatine, At_2, will react with sodium iodide, NaI.

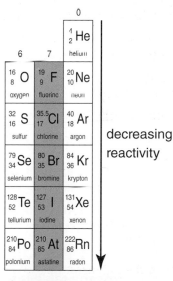

Figure 1.49

decreasing reactivity

KEY INFORMATION

Bromine displaces iodine from iodide solutions. It is seen as a red-brown solution.

$$Br_2 + 2KI \rightarrow 2KBr + I_2$$
(red-brown solution)

Reaction trends and predicting reactions

Learning objectives:

- explain why the trends down the group in Group 1 and in Group 7 are different
- explain the changes across a period
- predict the reactions of elements with water, dilute acid or oxygen from their position in the periodic table.

Fluorine, at the top of Group 7, is more reactive than iodine lower down the group. Yet why is caesium, lower down Group 1, so much more reactive with water than lithium at the top of the group? It is all to do with the electronic structure and the outer electrons.

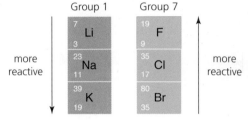

Figure 1.50 (a) Group 1; (b) Group 7

Opposite trends

The **trends** in **reactivity** in Group 1 and Group 7 are in opposite directions.

Reactivity increases down Group 1 as the outer electron in a potassium atom is further away from the nucleus than in a lithium atom so there is 'less pull' on it by the nucleus so it is lost *more* easily. So potassium is more reactive than lithium.

Reactivity decreases down Group 7 as the electron trying to transfer into a bromine atom is further away from the nucleus than in a fluorine atom so there is 'less pull' on it by the nucleus so it transfers in *less* easily. So bromine is less reactive than fluorine.

1 **Explain why the element caesium (also in Group 1) is much more reactive than sodium.**

2 **Explain why astatine is less reactive than chlorine.**

Trends across the table

The trend across the table is from metallic to non-metallic.

The trends across and down the periodic table depend on atomic structure and the **electronic structure.**

In Periods 2 and 3 across the table the outer electrons will increase by one from Group 1 with one outer electron and Group 2 with two outer electrons up to Group 7 with seven outer electrons.

Group 1 elements lose one electron to form positive ions easily. It is more difficult for elements to lose two or three electrons so they are less reactive. Group 7 elements gain one electron to form ions.

Group 1	Group 2											Group 3	Group 4	Group 5	Group 6	Group 7	0
			$^{1}_{1}$H hydrogen														$^{4}_{2}$He helium
$^{7}_{3}$Li lithium	$^{9}_{4}$Be beryllium											$^{11}_{5}$B boron	$^{12}_{6}$C carbon	$^{14}_{7}$N nitrogen	$^{16}_{8}$O oxygen	$^{19}_{9}$F fluorine	$^{20}_{10}$Ne neon
$^{23}_{11}$Na sodium	$^{24}_{12}$Mg magnesium											$^{27}_{13}$Al aluminium	$^{28}_{14}$Si silicon	$^{31}_{15}$P phosphorus	$^{32}_{16}$S sulfur	$^{35}_{17}$Cl chlorine	$^{40}_{18}$Ar argon
$^{39}_{19}$K potassium	$^{40}_{20}$Ca calcium	$^{45}_{21}$Sc scandium	$^{48}_{22}$Ti titanium	$^{51}_{23}$V vanadium	$^{52}_{24}$Cr chromium	$^{55}_{25}$Mn manganese	$^{56}_{26}$Fe iron	$^{59}_{27}$Co cobalt	$^{59}_{28}$Ni nickel	$^{64}_{29}$Cu copper	$^{65}_{30}$Zn zinc	$^{70}_{31}$Ga gallium	$^{73}_{32}$Ge germanium	$^{75}_{33}$As arsenic	$^{79}_{34}$Se selenium	$^{80}_{35}$Br bromine	$^{84}_{36}$Kr krypton

metals non-metals

Figure 1.51

3 **Explain why neon is an unreactive element.**

4 **Explain why sodium is a more reactive metal than aluminium.**

Predicting reactions

Knowing the position of an element in the periodic table will allow us to predict its behaviour with water, acid or oxygen.

From the trends that you know, make some predictions.

- Will barium be more reactive with dilute acid than magnesium?
- Will sulfur react more like phosphorus with oxygen or more like sodium?
- If bromine is bubbled through potassium chloride solution what reaction will there be?

Barium is lower down Group 2 than magnesium, so will lose electrons more easily, and so is more reactive.

Sulfur is more like phosphorus as they are both non-metals, so they will react with oxygen in a similar way.

There will be no reaction because chlorine is more reactive than bromine and so will not be displaced.

5 **Predict how rubidium will behave with water, acid and oxygen, giving reasons for your predictions.**

6 **Predict how strontium will react with dilute acid and give reasons for your predictions.**

7 **Hydrogen reacts with halogens to form hydrogen halides.**

a **Give the balanced chemical equation for the reaction of chlorine with hydrogen.**

b **Chlorine and hydrogen explode in sunlight. Predict the reactivity of fluorine with hydrogen. Explain your answer.**

KEY INFORMATION

An atom with one or two electrons will tend to lose them to make positive ions. For example, Na loses one electron to become Na^{+}.

MATHS SKILLS

Standard form and making estimates

Learning objectives:

- recognise the format of standard form
- convert decimals to standard form and vice versa
- make estimates without calculators so the answer in standard form seems reasonable.

Here, we're looking at using expressions in standard form and seeing how this helps us understand the size of very small entities such as atoms and ions.

When we talked earlier about an atom we used a model to describe it. We imagined it as a sphere with a radius of about 0.000 000 000 1 m. We also saw that the radius of the nucleus of the atom is about 0.000 000 000 000 01 m. It is very awkward to keep writing so many zeros, as it is easy to lose count and it is not so easy to see the comparison between one number and the other. Let's look at another way of writing these numbers using **standard form**.

Positive powers of ten for very large numbers

We write 1, 10 and 100 knowing what we mean. We can also write them as 1, 1×10 and $1 \times 10 \times 10$. We also know that 10×10 is 10^2. So 100 is 10^2. We can write the numbers as 1, 1×10 and 1×10^2.

1. Write 1 000 000 000 in standard form.

2. Write out the number 1×10^8

Standard form	M	HTh	TTh	Th	H	T	U	.	t
							1	.	0
1×10						1	0	.	0
1×10^2					1	0	0	.	0
1×10^3				1	0	0	0	.	0
1×10^4			1	0	0	0	0	.	0
1×10^5		1	0	0	0	0	0	.	0
1×10^6	1	0	0	0	0	0	0	.	0

10^6 is *not* 10 multiplied by itself 6 times. It is 10 multiplied by itself 5 times.

What about writing bigger numbers in standard form?

The **decimal point** is fixed and the position, or place value of the most significant digit, shows how big a number is.

To write 1 000 000 in standard form, take the first number on the left, which is 1. Looking at the table, how many places do we have to move the 1 to the right to reach the decimal point? We have to move the 1 six places to the right. So in standard form 1 000 000 is 1×10^6. The number 6 tells you how many tens there are when you write the number as a multiplication of 10 ($10 \times 10 \times 10 \times 10 \times 10 \times 10$).

Negative powers of ten for very small numbers

It is also possible to write numbers smaller than 1 in this form. 1 divided by 10 is 0.1. The number 1 has moved one place to the right of the decimal point. This is written as 1×10^{-1} in standard form. What is 1 divided by 100? The number 1 moves two places to the right of the decimal point to be 0.01. In standard form this is 1×10^{-2}.

standard form	U	.	t	h	th	Tth	Hth	milliionth
1×10^{-1}	0	.	1					
1×10^{-2}	0	.	0	1				
1×10^{-3}	0	.	0	0	1			
1×10^{-4}	0	.	0	0	0	1		
1×10^{-5}	0	.	0	0	0	0	1	
1×10^{-6}	0	.	0	0	0	0	0	1

Converting numbers to standard form

Standard form can also be used to represent numbers where the most significant digit is not 1. For example, the ordinary number 6000 can be written as 6×1000, or 6×10^3 in standard form.

Remember that standard form always has exactly one digit bigger than or equal to 1 but less than 10. 0.3×10^4 is not in standard form. It is 3×10^3 in standard form.

Some big and small numbers that you have already met in Chemistry are:

Avogadro's number	Atomic radius (m)	Nuclear radius (m)	Mass of a gold atom (g)
6.023×10^{23}	1×10^{-10}	1×10^{-14}	3.3×10^{-22}

When you calculate with big and small numbers using a calculator it is essential that you first estimate what your answer should look like. Making an estimate of the result of the calculation can save you from making mistakes with your calculator. The best way to estimate the answer without a calculator is to round the numbers sensibly and then carry out the calculation in your head.

3 Write 0.000 000 000 000 000 001 in standard form.

4 Write out the number 1×10^{-9}

5 Calculate:
a $6 \times 10^9 \times 3 \times 10^3$
b $6 \times 10^9 \times 4 \times 10^{-2}$
c $6 \times \frac{10^8}{2} \times 10^2$

6 If you were able to lay the Avogadro's number of atoms in a straight line next to each other, how far would they stretch?

7 Calculate the mass of 3.0×10^{26} gold atoms using the mass of a single gold atom given in the data table.

KEY INFORMATION

To multiply two numbers in standard form you simply add the indices or powers of the tens. For example, $2 \times 10^{15} \times 3 \times 10^9$ is 2×3 with 10^{15+9}, which is 6×10^{24}. With smaller numbers, $2 \times 10^{-15} \times 3 \times 10^{-9}$ is 6×10^{-24}.

Check your progress

You should be able to:

name compounds from their formula → recall the names of the first 20 elements in the periodic table and the elements in Groups 1 and 7 → use symbol equations to describe chemical reactions

describe how to separate mixtures of elements and compounds → use word equations to describe chemical reactions → use balanced equations to describe reactions

explain that early models of the atom did not have shells with electrons → explain that early models of atoms developed as new evidence became available → explain why the scattering experiment led to a change in the atomic model

draw a diagram of a small nucleus containing protons and neutrons with orbiting electrons at a distance → calculate the numbers of sub-atomic particles in ions and isotopes given the atomic and mass numbers → complete data tables showing the atomic numbers, mass numbers and numbers of sub-atomic particles from symbols

describe how Mendeleev was able to leave spaces for elements that had not yet been discovered → explain why the modern periodic table has the elements in order of atomic number → explain how Mendeleev was able to make predictions of as-yet undiscovered elements such as eka-silicon

describe the pattern of the electrons in shells for the first 20 elements → explain how the electronic arrangement of atoms follows a pattern up to the atomic number 20 → explain how the electronic arrangements of transition metal atoms put them into a period

describe a number of physical properties of metals and non-metals → explain that atoms of metals have 1, 2 or 3 electrons in their outer shell → explain that non-metals need to gain or share electrons during reactions and that metals need to lose electrons during reactions

explain that non-metals are on the right-hand side of the periodic table → explain that non-metals have 4, 5 ,6, 7 or 8 electrons in their outer shell → predict the relative reactivity across the periods and give reasons

describe the unreactivity of the noble gases → explain the trend down Group 0 of increasing boiling point → explain the trend down Group 0 of increasing boiling point in terms of atomic mass

predict the reactions with water of Group 1 elements lower than potassium → predict and explain the relative reactivity down the groups → explain the trend down the group of increasing reactivity by electron structure

recall the colours of the halogens and the order of reactivity of chlorine, bromine and iodine → describe the order of reactivity and explain the displacement of halogens → predict displacement reaction outcomes of halogens other than chlorine, bromine and iodine

explain that a stable outer shell of electrons makes noble gases unreactive → predict the properties of 'unknown' elements from their position in the group → explain the trend of increasing reactivity in terms of electron structure

Worked example

Sam and Alex are researching some properties of Group 1 metals.

1 **Shade the section of the periodic table where the Group 1 metals are found.**

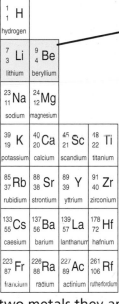

This is incorrect. The first column needs to be shaded.

The two metals they are researching are sodium and potassium.

2 **Write down two properties that these metals have.**

They are shiny when cut.

They have a very high density.

The first property is correct. However, sodium and potassium float on water so have a density less than water. The student may be confusing Group 1 metals with transition metals, which have high density.

Sam and Alex find out that sodium and potassium react with water. They find that sodium reacts with water to make sodium hydroxide and that hydrogen is given off.

3 **Write a word equation for the reaction.**

sodium + water → sodium hydroxide

The reactants are correct but hydrogen needs to be written as a product on the right-hand side.

Sam says that potassium reacts more vigorously than sodium but Alex says that they are in the same group so they react the same.

4 **a** **Explain why Sam is correct about the trend.**

The lower down the group the better they react.

This answer could be expressed more clearly by substituting the word 'better' with 'more vigorously'.

b **Explain why Sam is correct using ideas about the structure of atoms.**

The bigger the atom the quicker the reaction.

This answer needs more detail. The further away the outer electron is from the nucleus the more easily it is 'lost', as the pull by the positive nucleus on the negative electron is less.

End of chapter questions

Getting started

1 Identify the gas given off when sodium reacts with water. `1 Mark`

2 Which of the following is a noble gas? `1 Mark`

   a oxygen
   b helium
   c chlorine
   d nitrogen

3 Suggest two differences between a metal and a non-metal. `2 Marks`

4 An atom has 3 protons, 4 neutrons and 3 electrons. What is its atomic mass? `1 Mark`

   a 3
   b 6
   c 7
   d 10

5 Determine the electron arrangement in an atom with 10 electrons. `1 Mark`

6 Explain why an atom with an electron pattern of 2,8,1 is in Group 1 and Period 3 of the periodic table. `2 Marks`

7 Jo and Gita carefully observe three halogens. Match their observation to the correct halogen.

Match the flame colour that they see with the correct metal of the compound. `2 Marks`

orange liquid	iodine
pale green gas	bromine
grey solid	chlorine

Going further

8 When sodium reacts with water, hydrogen is given off. Identify the other product. `1 Mark`

9 To which group does chlorine belong? `1 Mark`

   a 0
   b 1
   c 6
   d 7

10 Describe the trend of reactivity of the Group 1 metals. `2 Marks`

11 Explain where the atom with an electron pattern of 2,8,2 is positioned in the periodic table. Suggest why it is a metal. `4 Marks`

12 Halogen X has a boiling point of 59 °C and halogen Y a boiling point of 184 °C. `2 Marks`

   a Justify which halogen has the lower atomic number.
   b Explain which halogen is more reactive.

More challenging

13 Identify the number of electrons in an atom of $^{31}_{15}$P.

1 Mark

14 There are two atoms, $^{28}_{14}$Si and $^{30}_{14}$Si.

Work out how many neutrons each atom contains.

1 Mark

15 Explain why the reactivity of Group 1 metals increases down the group.

2 Marks

16 Francium is a highly radioactive and rare alkali metal.

2 Marks

	Melting point / °C	Density / g cm^{-3}
Rb	39.3	1.53
Cs	28.4	1.93
Fr	D	E

Predict the melting point, D, and density, E, of francium using the data in the table.

17 Chlorine (Cl_2) is bubbled into a solution of potassium iodide (KI).

  a Describe the reaction.
  b Explain why this reaction is able to take place.

4 Marks

Most demanding

18 Deduce the number of the sub-atomic particles that make up the atom $^{31}_{15}$P.

2 Marks

19 Explain why the atom with an electron pattern of 2,8,6 is a non-metal. Explain why this atom is less reactive than the atom with an electron pattern of 2,6 or 2,7.

4 Marks

20 The table shows the data and properties for an element.

Atomic number	Electron pattern	Reaction with water
37	2,8,8,18,1	Violently, giving off hydrogen and forming an alkali

Deduce which type of element it is, the group it belongs to and why it reacts so violently with water.

4 Marks

Total: 40 Marks

STRUCTURE, BONDING AND THE PROPERTIES OF MATTER

IDEAS YOU HAVE MET BEFORE:

STATES OF MATTER AND PARTICLE MODEL

- Ice and other solids can turn to liquids and gases.
- Solids turn into liquids at the melting point.
- Liquids turn into gases at the boiling point.

ENERGY TRANSFERS AND CHEMICAL BONDS

- To melt salt takes more energy than to melt ice.
- Oxygen and nitrogen boil at very low temperatures.
- Elements combine by making chemical bonds.

DIFFERENT KINDS OF CHEMICAL BONDS

- Salt dissolves in water but sand does not.
- Metals conduct electricity.
- Plastics do not conduct electricity.

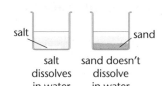

salt   sand

salt dissolves in water   sand doesn't dissolve in water

METALS AND ALLOYS

- Metals are difficult to cut.
- Metals can be melted and cast into statues and bells.
- Metals can be hammered and drawn into wires.

STRUCTURE AND BONDING OF CARBON

- Carbon is an element that we know as soot and charcoal.
- Diamonds are hard, precious gems that are cut to sparkle.
- Graphite is used in pencil leads and as electrodes.

IN THIS CHAPTER YOU WILL FIND OUT ABOUT:

WHAT HAPPENS TO PARTICLES AS SUBSTANCES CHANGE STATE?

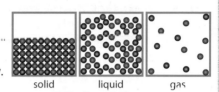

- Particles of solids are arranged regularly and pack closely.
- Particles of liquids and gases move more freely and rapidly.
- Particle models use 'solid' spheres but particles are not solid.

WHY IS SO MUCH ENERGY NEEDED TO MELT SOME SUBSTANCES?

- Silicon dioxide and diamonds have bonds in all directions.
- Breaking chemical bonds requires a great deal of energy.
- Most metals are difficult to melt due to strong bonding.

ARE THERE DIFFERENT TYPES OF CHEMICAL BONDS?

- Sodium chloride is bonded by ionic bonding.
- Diamond is bonded by covalent bonding.
- Metals are bonded by metallic bonding.

WHY CAN METALS CONDUCT ELECTRICITY?

- Metallic bonding uses delocalised electrons.
- Delocalised electrons can move freely through the metal.
- Electric current is electron flow.

WHY ARE DIAMONDS SO HARD AND GRAPHITE SO SOFT?

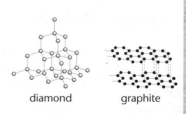

diamond   graphite

- Diamond is a giant structure with tetrahedral bonds.
- Graphite is made of hexagonal rings of carbon atoms in layers.
- There are weak forces between the layers in graphite, meaning that the layers can slide over each other.

Chemical bonds

Learning objectives:

- describe the three main types of bonding
- explain how electrons are used in the three types of bonding
- explain how bonding and properties are linked.

KEY WORDS
....
covalent
delocalised
ionic
metallic

Why do sodium chloride, carbon dioxide and copper look so different and behave so differently? It is all to do with the number of electrons in each of the atoms that make up the substances and how these electrons move. They may move in a localised arrangement or in a delocalised arrangement.

Types of bonds

There are three types of strong chemical bonds:

ionic   **covalent**   **metallic.**

	Ionic	Covalent	Metallic
This bonding occurs in:	compounds formed from metals combined with non-metals	compounds of non-metals and in most non-metallic elements	metallic elements and alloys
In this bonding the particles are:	oppositely charged ions	atoms that share pairs of electrons	atoms that share delocalised electrons

1 Compare ionic and covalent bonding.

How do atoms join together?

Atoms can join together by *transferring* electrons or by *sharing* electrons – and some atoms do both.

When metals and non-metals *transfer* electrons the atoms become charged. These charged atoms are called *ions*. The metal ions are positively charged. The non-metal ions are negatively charged. The positive ions and negative ions are

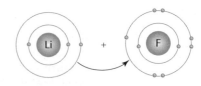

Figure 2.1

attracted to each other and bond together by electrostatic attraction. This bonding is known as *ionic bonding.*

Non-metals can also *share* electrons between atoms. This type of bonding is known as *covalent bonding*.

Metals form positive ions which are attracted to the negatively charged delocalised electrons that surround them. This is called *metallic bonding*.

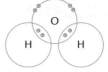

Figure 2.2

These three types of bonding produce different kinds of substances.

	Bonding		
	Ionic	Covalent	Metallic
Structures are:	large crystals made from ions attracted to each other by electrostatic attraction	molecules made from atoms bonded by covalent bonds	lumps or sheets of metal made from atoms packed together so that the delocalised electrons move through the fixed positive ions
Examples are:	sodium chloride and magnesium oxide	carbon dioxide and water	copper and gold
At room temperature they are:	solids	gases and liquids	solids (except for mercury which is a liquid at room temperature)
Melting points are:	high	low	high
Do they conduct electricity?	Not as a solid, only when melted or in solution.	Not at all or poorly.	Yes, good conductors.

2 Substance **R** is a good electrical conductor as a solid and liquid and has a high melting point. Suggest the type of bonding.

3 Describe what a delocalised electron does in a metal.

4 Suggest why covalently bonded substances such as carbon dioxide cannot conduct electricity.

Why do atoms form bonds?

Atoms with an outer shell of eight electrons have a stable electronic structure. These atoms are unreactive. Examples of such unreactive atoms are the elements neon, argon and krypton (all from group 0).

However, most atoms usually have too many or too few electrons in their outer shell to be stable and will need to gain or lose electrons.

It is easiest for an atom that has one or two electrons in its outer shell to lose them. These are metal atoms which form positive ions.

It is easiest for an atom that has one or two spaces in its outer shell to fill to gain electrons. There are two ways this can happen: non-metal atoms can form negative ions and non-metal atoms can also share electrons.

5 Explain why it is the *electrons* that are involved in bonding.

DID YOU KNOW?

Metals conduct electricity because there are delocalised electrons.

KEY INFORMATION

Delocalised electrons are often called a 'sea' of electrons in metals.

Ionic bonding

Learning objectives:

- represent an ionic bond with a diagram
- draw dot and cross diagrams for ionic compounds
- work out the charge on the ions of metals and non-metals from the group number of the element (1, 2, 6 and 7).

Electron transfer between atoms can be drawn in many ways. One way is by drawing electron patterns in shells and using arrows to show the transfer. A conventional way is to draw a 'dot and cross' diagram. Diagrams help us to visualise what is happening at a level that we cannot see even with the most powerful microscope.

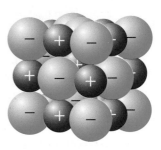

Figure 2.3 An electron is transferred *from* a metal atom *to* a non-metal atom.

Making ions

Noble gases (Group 0 of the periodic table) have stable atoms. Except for helium, their outer shell of electrons contains eight electrons.

A metal atom has a small number of electrons in its outer shell.	A non-metal atom has spaces in its outer shell.
It needs to lose these electrons to attain a stable electronic structure.	It needs to gain electrons to be stable.
The electrons transfer *out* from the metal atom to make a stable ion.	The electrons transfer *in* to the non-metal atom to make a stable ion.

The electrons transfer *out from* a metal atom *in to* a non-metal atom.

A metal atom becomes a positive ion.	A non-metal atom becomes a negative ion.

The positive ions and the negative ions are then attracted to one another. These attractions spread to a number of other ions to make a solid lattice.

Figure 2.4 Ions of opposite charge are attracted together as a solid lattice.

1 Explain how a metal atom becomes a positive ion.

2 Explain how solid ionic lattices are formed from elements, using sodium chloride as an example.

DID YOU KNOW?

The charge on the positive ion corresponds to the group number for Group 1 and 2 ions.

Dot and cross diagrams

When a metal atom reacts with a non-metal atom, electrons in the outer shell of the metal atom are transferred.

Metal atoms lose electrons to become positively charged ions.	Non-metal atoms gain electrons to become negatively charged ions.
The ions produced by metals in Groups 1 and 2 have the electronic structure of a noble gas (Group 0) and a positive charge.	The ions produced by non-metals in Groups 6 and 7 have the electronic structure of a noble gas (Group 0) and a negative charge.

KEY INFORMATION

In dot and cross diagrams only the outer electrons are represented. No inner shells are shown. The ions are drawn with **square brackets** and the charge on the outside top right.

The **electron transfer** during the formation of an ionic compound can be represented by a **dot and cross** diagram. For example, for sodium chloride.

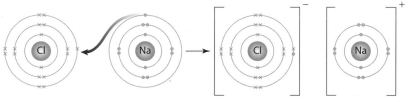

Full shell model

Figure 2.5

As shown here, electrons in dot and cross diagrams are sometimes represented in just the outermost shell since this determines the atom's bonding capabilities.

Sodium atoms have the electron pattern (2,8,1) and they have 1 outer electron.	Chlorine atoms have the electron pattern (2,8,7) and they have 7 outer electrons.
This electron is represented by 1 dot.	These electrons are represented by 7 crosses.

When the ions form, the 'dot' transfers out to the space in amongst the 'crosses'.

The sodium ion has an electron pattern (2,8) and is positively charged overall.	The chloride ion has an electron pattern (2,8,8) and is negatively charged overall.

Figure 2.6 Ions held together forming crystals.

- Both ions have a stable electron pattern.
- The ions are attracted together by the **electrostatic attraction** of opposite charges.
- They attract to make large packs of ions that form crystals.

3 State the electron arrangement of a stable neon atom.

4 Predict how many electrons a fluorine atom needs to become an ion.

5 Explain why a chloride ion is negative. (Hint: use ideas about the nucleus of an atom.)

Ionic bonds

If an atom loses electrons a positive ion is formed. This is because there will be fewer negatively charged electrons than the number of positively charged protons in the nucleus.

If an atom gains electrons a negative ion is formed. This is because there will be more negatively charged electrons than the number of positively charged protons in the nucleus.

If magnesium makes a 2+ ion and chlorine makes a 1– ion, then *two* chlorine ions are needed to bond to *one* magnesium ion to make $MgCl_2$. See Figure 2.8.

6 Draw an electron transfer diagram to show the bonding of sodium oxide.

DID YOU KNOW?

When a Group 1 atom loses one electron a (positive) 1+ ion is formed, for example, $Na \rightarrow Na^+$.

When a Group 2 atom loses two electrons a (positive) 2+ ion is formed, for example, $Mg \rightarrow Mg^{2+}$.

When a Group 7 atom gains one electron a (negative) 1– ion is formed, for example, $F \rightarrow F^-$. See Figure 2.7.

When a Group 6 atom gains two electrons a (negative) 2– ion is formed, for example, $O \rightarrow O^{2-}$.

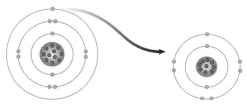

Figure 2.7

Figure 2.8

Ionic compounds

Learning objectives:

- identify ionic compounds from structures
- explain the limitations of diagrams and models
- work out the empirical formula of an ionic compound.

Positive ions and negative ions attract each other in all directions to form giant ionic lattice **structures with the electrostatic forces acting in all directions. Models and diagrams help us to understand this type of bonding and the structures that are formed.**

Giant structures

Ionic bonding occurs when oppositely charged ions are held together by strong **electrostatic** forces of attraction.

An ionic compound is a giant structure of ions in a lattice. One example of an ionic compound is sodium chloride.

The structure of sodium chloride can be demonstrated by different **3D representations** (Figure 2.9)

(a)   (b)

Key
● Na^+
○ Cl^-

Figure 2.9 Models of NaCl
(a) ball and stick (b) close packed

	Type of diagram:	
	ball and stick (a)	close packed (b)
The diagram is helpful to show:	the structure in 3D the charges on the ions the arrangement of ions in 3D the *type* of ions in all directions	the structure in 3D the charges on the ions the arrangement of ions in 2D (look at the front face of the diagram) the closeness of ions
The diagram is unhelpful:	the ions are actually close together gives a false image of bond direction when it is only electrostatic attraction	difficult to see the arrangement of ions in 3D
This diagram is best to represent:	the number and type of ions in 3D	the way that ions are packed close together

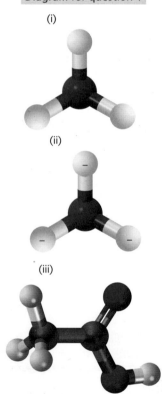

Diagram for question 1

(i)

(ii)

(iii)

1 **Which one of the three models shows an ionic compound? i, ii or iii?**

Empirical formula

The **empirical formula** is the simplest ratio of ions. In magnesium oxide the ions are held together by electrostatic attraction, to form a giant structure, in exactly the same way

DID YOU KNOW?

In an ionic compound each atom has either gained or lost electrons to achieve a stable electronic structure.

as in sodium chloride, except that the magnesium ion has a 2+ charge and the oxide ion has a 2– charge.

Sodium chloride has the ratio of one Na^+ ion to one Cl^- ion. The empirical formula for sodium chloride is NaCl.

Magnesium oxide has a ratio of one Mg^{2+} ion to one O^{2-} ion. The empirical formula for magensium oxide is MgO.

Other compounds that bond using ionic bonds are sodium oxide and magnesium chloride.

(a) dot and cross diagram

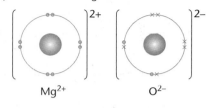

Mg^{2+} O^{2-}

(b) 3D close packed diagram

Figure 2.10 The bonding in magnesium oxide

Sodium only has one electron to lose, but oxygen needs to gain two electrons. Two sodium atoms are needed to bond with one oxygen atom.	Magnesium needs to lose two electrons, but chlorine can only gain one electron. So one magnesium atom needs two chlorine atoms to achieve an ionic bond and make magnesium chloride.

(a) 2D electron transfer diagram of sodium oxide

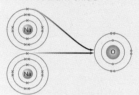

(a) 2D electron transfer diagram of magnesium chloride

(b) Ball and stick diagram of sodium oxide

(b) Ball and stick diagram of magnesium chloride

Diagram for question 3

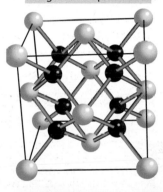

2 Potassium fluoride is formed from K^+ and F^- ions. Show the ionic bonding in potassium fluoride using an electron transfer diagram and a dot and cross diagram.

3 Describe why, in terms of electrons, potassium sulfide (shown in the diagram to the right) has an empirical formula of K_2S.

KEY INFORMATION

The ions at the corners are shared by four cubes, so only count as 1/4 and the ions in the faces are shared by two cubes and so only count as 1/2. How many yellow ions are on corners? How many yellow ions are on faces of the cube? Are all the purple ions inside the cube?

HIGHER TIER ONLY

Equations of ion formation

$Na - e^- \rightarrow Na^+$	$Cl + e^- \rightarrow Cl^-$
$Mg - 2e^- \rightarrow Mg^{2+}$	$O + 2e^- \rightarrow O^{2-}$

To make $MgCl_2$:

$$Mg - 2e^- \rightarrow Mg^{2+}$$

$$2Cl + 2e^- \rightarrow 2Cl^- \text{ Empirical formula } MgCl_2$$

4 Draw a dot and cross diagram to show the formation of calcium chloride and write its empirical formula.

5 Determine the empirical formula of the following ionic compounds given the ions: **a** Na^+ and N^{3-} **b** Al^{3+} and O^{2-}.

REMEMBER!

Mg loses 2 electron.
O gains 2 electrons.

Covalent bonding

Learning objectives:

- recognise substances made of small molecules from their formula
- draw dot and cross diagrams for small molecules
- deduce molecular formulae from models and diagrams.

KEY WORDS

covalent
giant covalent
  structure
polymer
single bond

When two or more non-metal atoms bond together they *share* electrons. In this way they form small *individual* molecules. Such molecules can exist on their own, and don't have to be part of a large structure. They are freer to move around so the collection of molecules exists as liquids or gases.

Sharing electrons

When two or more non-metal atoms bond together they form a molecule. Molecules are bonded by **covalent** bonds. These bonds are very strong. An example is a hydrogen molecule, H_2. Each hydrogen atom has a single electron so the two atoms *share* their electrons as a pair to have the stability of a full outer shell.

This can be represented as:

(a) (H ⨯ H)   (b) H–H   (c) H_2

Only one pair of electrons is shared so it is a **single covalent bond**.

1 **Hydrogen and fluorine bond with a single covalent bond. Draw a diagram to show this.**

Covalently bonded molecules

Other small molecules that are covalently bonded are hydrogen chloride, chlorine, water, ammonia and methane. These are bonded by single bonds when one pair of electrons is shared.

> **DID YOU KNOW?**
>
> Some covalently bonded substances may consist of small molecules, but some covalently bonded substances have very large molecules, such as **polymers** (topic 2.9). Some covalently bonded substances have **giant covalent structures**, such as as diamond and silicon dioxide (topic 2.10).

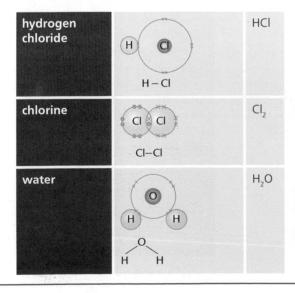

hydrogen chloride		HCl
chlorine		Cl_2
water		H_2O

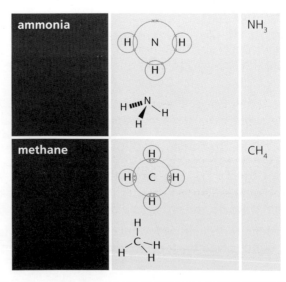

| ammonia | | NH_3 |
| methane | | CH_4 |

These diagrams in 2D have their limitations as they do not show the 3D shape of the molecule.

Covalent bonds act in a particular direction. 3D ball and stick models or space-filling models can provide a better indication of the structures.

When atoms form covalent bonds they share pairs of electrons. This can be shown using a dot and cross diagram.

The outer electrons are written as pairs of electrons. The bond pair is written between the two symbols.

hydrogen	hydrogen chloride	chlorine
H H	H Cl	Cl Cl

water	ammonia	methane
O H H	H N H H	H C H H H

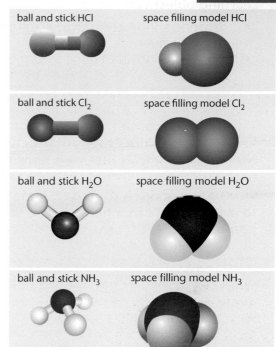

ball and stick HCl · space filling model HCl

ball and stick Cl₂ · space filling model Cl₂

ball and stick H₂O · space filling model H₂O

ball and stick NH₃ · space filling model NH₃

ball and stick CH₄ · space filling model CH₄

2 The displayed formula of hydrogen fluoride is H–F. Represent HF as a dot and cross diagram.

3 How many bond pairs are there in a molecule of ammonia?

4 Determine the molecular formula of methane.

5 Draw a dot and cross diagram for H_2S.

Multiple bonds

One 'dot and cross' together represents a single bond, which is one bond pair of electrons. Two pairs of 'dots and crosses' side by side represent a double bond (as in O_2). Three electron bond pairs is a triple bond (as in N_2).

6 Explain why there are two dots and two crosses between the oxygen atoms in the diagram to the right.

7 Use the dot and cross diagram to the right to explain why a molecule of nitrogen N_2 has a triple bond.

8 Draw a dot and cross diagram to show the bonding in carbon dioxide CO_2. (Hint: it is a linear molecule with a C atom in the middle.)

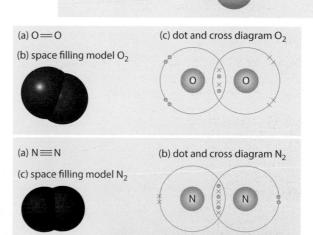

(a) O=O
(b) space filling model O_2
(c) dot and cross diagram O_2

(a) N≡N
(b) dot and cross diagram N_2
(c) space filling model N_2

KEY INFORMATION

The inner shells of electrons are not shown. There is no charge needed after the square bracket as the molecule is neutral.

Metallic bonding

Learning objectives:

- describe that metals form giant structures
- explain how metal ions are held together
- explain delocalisation of electrons.

KEY WORDS

delocalised
    electrons
metal ions
metallic bonding
'sea' of electrons

Think about copper, iron and gold. What is it about metals that makes them distinctive? Metals conduct electricity and can often be hammered into sheets, bent and twisted. This is because of the way that metal atoms are bonded together. Metallic bonds are strong. Metallic bonds involve the free movement of electrons through the giant structure.

Giant structures

Metals consist of giant structures of atoms arranged in a regular pattern.

The metal atoms have a small number of electrons in their outer shell, which are free to move among the atoms.

The electrons move through the whole structure making the bonds between the atoms strong.

These moving electrons are described as delocalised **electrons**. As they leave the atoms to move they leave behind a positive ion.

So the diagram of **metallic bonding** is drawn as:

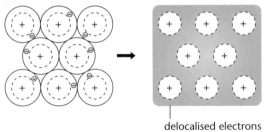

delocalised electrons

Figure 2.11 Metallic bonding

① **Metals have a small number of electrons in their outer shell which are free to move. Show this by describing the electron pattern in magnesium.**

② **Describe metallic bonding using lithium as an example.**

Holding a metal together

A metal is made of atoms that are close packed and held together by the delocalised electrons.

The metallic bond is the strong electrostatic force of attraction between the positive **metal ions** and the **delocalised electrons**.

DID YOU KNOW?

We talk about *electrostatic attraction* for electrical charge that is built up but slow to move and *conduction of electricity* for movement of charge. You know about static electricity from when you rubbed balloons with woollen cloth and could stick them on a wall.

Figure 2.12 Close-packed atoms of a metal

The moving electrons are not localised to a particular atom but are *delocalised* and free to move through the whole structure.

The sharing of the **delocalised electrons** makes, strong *metallic bonds*.

If the metals need to be melted then the metallic bonds need to be disrupted. This is done by heating the metal, so giving the positive ions more energy so that they move with more energy. The electrons also move with more energy. In general the metals that only have one spare electron to contribute to the metallic bond melt at a lower temperature than those which donate two to the bond.

The delocalised electrons are the key factor in a metal's ability to conduct electricity. An electrical current is the movement of these delocalised electrons through the lattice of ions.

All metals have this property, although some conduct electricity better than others. The higher the number of delocalised electrons a metal produces, the better it's ability to conduct electricity.

KEY INFORMATION

This explanation is not the whole story, though. There are other factors involved, such as lattice structures, so be careful when making predictions.

KEY INFORMATION

Remember why metals form positive ions. Look at where metals are found in the periodic table and remember the electron pattern for metals.

Metal	lithium	calcium	aluminium	copper	silver
Relative electrical conductivity	1.08	2.98	3.50	5.96	6.30

3 Explain the term delocalised electron.

4 Explain why silver is such a good conductor but **solid** silver oxide is not.

Consequences of electron delocalisation

A metal conducts electricity because delocalised electrons within its structure can move easily through it.

Metals often have high melting points and boiling points. This is because a lot of energy is needed to overcome the strong attraction between the delocalised electrons and the positive metal ions.

5 Support the theory of the structure of metals by explaining how a metal expands when it is heated.

6 Suggest why aluminium is a better electrical conductor than sodium.

Chemistry

Three states of matter

Learning objectives:

- use data to predict the states of substances
- explain the changes of state
- use state symbols in chemical equations.

KEY WORDS

changes of state
condensing
limitations
particle

What happens to ice in a cold drink? How can we describe what is happening as the solid becomes a liquid? A particle model can be used to describe the three states of matter – solid, liquid and gas – and to help us to visualise what happens during changes of state.

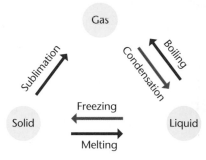

Figure 2.13 The three states of matter

The states of matter

The three states of matter are solid (s), liquid (l) and gas (g).

- Melting and freezing take place at the melting point.
- Boiling and **condensing** take place at the boiling point.

Substance	Melting point (°C)	Boiling point (°C)	State at room temperature
W	−18	42	liquid
X	150	875	
Y	−190	−84	
Z	−56	16	

Substance W in the table will have melted at −18 °C but at room temperature, which is 25 °C, it will not have boiled. So it is a liquid.

The three states of matter can be represented by a simple model. In this model, **particles** are represented by small solid spheres.

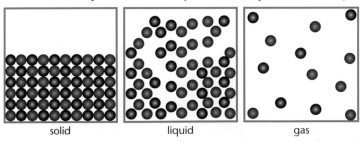

Figure 2.15 Particle model diagrams of solid, liquid and gas

Look again at substance W in the table. At 25 °C it will be a liquid. However, a few particles will have enough energy to escape the surface of the liquid and evaporation will be taking place.

DID YOU KNOW?

Iodine can turn straight from a solid to a gas. This is called sublimation.

Figure 2.14 Iodine changing from a solid to a gas

1 **What are the states of substances X, Y and Z at room temperature in the table?**

Changes of state

Particle theory can help to explain the **changes of state**. Melting, boiling, freezing and condensing are processes that depend on changing the forces *between* the **particles**.

In melting and boiling the strength of the forces between particles decreases. The distance between particles increases and the arrangement becomes more random. The particles move more so more energy is needed from the surroundings.	In freezing and condensing the strength of the forces between the particles remains the same. The distance between particles decreases and the arrangement become less random. The particles move less so less energy is needed.

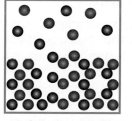

Figure 2.16 Particle model of evaporation

The *amount* of energy needed to change state from solid to liquid and from liquid to gas depends on the strength of the forces between the particles of the substance.

The strength of the forces between particles depends on the nature of the particles involved, on the type of bonding and on the structure of the substance.

The stronger the forces between the particles are, the higher the melting point and boiling point of the substance.

2. Look at the data for substance W in the table. Describe what is happening to the particles and the forces between them at these temperatures: −28 °C, −18 °C, −14 °C, 25 °C, 38 °C, 42 °C, 46 °C

3. Identify the states of the substances in these equations:

   a $KI(aq) + AgNO_3(aq) \longrightarrow KNO_3(aq) + AgI(s)$

   b $ZnCO_3(s) \xrightarrow{\text{heat}} ZnO(s) + CO_2(g)$

4. Suggest why iron has high melting and boiling points.

5. Ethanol has a boiling point of 78.4 °C. Propane has a boiling point of −42 °C. Suggest why.

> **KEY INFORMATION**
>
> These diagrams are only diagrams that show a model. For example, in the model of a solid the individual spheres represent particles that are not, themselves, solids. It is only when lots of these particles are arranged closely in a regular pattern that they together represent a solid.

> **DID YOU KNOW?**
>
> In chemical equations, the three states of matter are shown as (s), (l) and (g), with (aq) for aqueous solutions.

HIGHER TIER ONLY

Limitations of models

This simple model is **limited** as:

- there are no forces represented between the spheres
- all the particles are represented as spheres
- the spheres are represented as solid and inelastic.

This means that the change in forces and collisions between particles cannot be represented fully. However, it is a useful model to show spatial arrangement, both regular and random.

6. Explain the limitations of the particle theory when considering the process of condensing.

Properties of ionic compounds

Learning objectives:

- describe the properties of ionic compounds
- relate their melting points to forces between ions
- explain when ionic compounds can conduct electricity.

KEY WORDS

conducting electricity
energy
high melting point
ions free to move

Ionic compounds are giant structures that require a lot of energy to break down. As solids they do not conduct electricity. However, they do conduct electricity under certain conditions, once their giant lattice is broken down.

Properties of ionic compounds

Ionic compounds, such as sodium chloride, have regular structures. They make giant ionic lattices of oppositely charged ions. These ions have strong electrostatic forces of attraction in all directions.

These compounds:

have **high melting points** and high boiling points	because	the forces of attraction holding the ions together are very strong
conduct electricity when melted or dissolved in water	because	the **ions are free to move** and so charge can flow

The forces of attraction have to be overcome for sodium chloride to melt so a great deal of **energy** is required to break the lattice down to individual, moving, ions.

1 The melting point of sodium chloride is 804 °C. Why does sodium chloride have a high melting point?

Why do ionic compounds have these properties?

These compounds have high melting points and high boiling points because of the large amounts of energy needed to break the many strong ionic bonds.

The ionic compounds have giant ionic lattices in which there are strong electrostatic forces of attraction in all directions between oppositely charged ions.

These forces have to be overcome for the particles to move freely and more randomly as a liquid or a gas.

When melted or dissolved in water, ionic compounds can conduct electricity.

This is because ionic compounds, such as sodium chloride, are made of ions that become free to move when melted or dissolved.

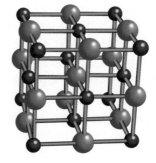

Figure 2.17 There is a strong attraction between the positive and negative ions of sodium chloride, forming a giant lattice in the shape of a cube.

DID YOU KNOW?

Each magnesium atom donates two electrons to the oxygen atom, which makes a stronger ionic bond than the bonds made by sodium transferring one electron to chlorine atoms.

Ionic compounds do conduct electricity when they are melted (molten) because the ions are free to move and so charge can flow between electrodes.

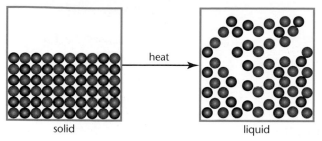

solid   liquid

Figure 2.18 Energy is needed to overcome the strong attraction between positive and negative ions of ionic compounds, for the solid to become a liquid when melting.

Ionic compounds such as sodium chloride can dissolve in water to make a solution. Sodium chloride solution can conduct electricity as the ions are free to move around and carry charge to each electrode. Two oppositely charged electrodes are put into the solution to allow the electricity to pass through.

2 **Substance D has a high melting point and conducts electricity in solid and liquid state. Explain whether D is ionic.**

3 **Explain why sodium chloride does not conduct electricity when it is a solid.**

Comparing NaCl with other ionic compounds

Sodium chloride NaCl is a typical ionic compound. It dissolves in water, does not conduct electricity when solid, but does when melted or dissolved to form a solution, and it has a high melting point (804 °C).

Magnesium oxide MgO and potassium chloride KCl are other ionic solids. Look at the data table to see which solid will require the most energy to overcome the forces between its ions.

	Melting point °C	Conducts electricity when solid	Conducts electricity when molten
sodium chloride	804	No	Yes
magnesium oxide	2800	No	Yes
potassium chloride	770	No	Yes

These melting points are high because in all the compounds the ions form giant lattice structures that need a lot of energy to break down.

A lot more energy is needed to overcome the forces in MgO than in NaCl and KCl.

Can we explain why this should be?

4 **Suggest why magnesium oxide has a higher melting point than sodium chloride or potassium chloride.**

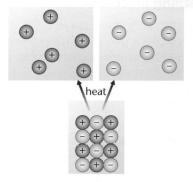

Figure 2.19 Ionic solid melting to produce ions that are free to move.

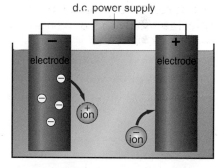

Figure 2.20 Sodium chloride solution can conduct electricity as the ions are free to move.

KEY INFORMATION

You only need to know the details about sodium chloride. In this example we are using evidence to make a reasoned suggestion. You do not need to know the details of magnesium oxide or potassium chloride.

Properties of small molecules

KEY WORDS

covalent bond
free electrons
intermolecular force
molecule

Learning objectives:

* identify small molecules from formulae
* explain the strength of covalent bonds
* relate the intermolecular forces to the bulk properties of a substance.

Researchers have found that the polar caps of Mars are probably made of both water ice and dry ice from carbon dioxide. However, at room temperature on Earth, carbon dioxide is a gas and water is a liquid. We meet them as bulk substances made up of small molecules and their properties are related to the size of their molecules.

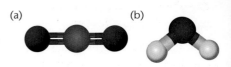

Figure 2.21 Models of small molecules (a) carbon dioxide (b) water

Small molecules

Carbon dioxide and water are examples of simple, small **molecules**.

You can often tell a small molecule from its formula. It has two or three atoms joined together, with no charge.

Atoms	O	Mg	Cl	Na	S
Ions	O^{2-}	Mg^{2+}	Cl^-	Na^+	SO_4^{2-}
Molecules	H_2	Cl_2	CO_2	Br_2	H_2O

KEY INFORMATION

The subscript tells you how many atoms of the same type are bonded. For example, H_2O means there are two hydrogen atoms bonded to one oxygen atom in one molecule of water.

Look at the table and notice that the molecules have no charge and more than one atom.

Substances that are made up of small molecules are usually gases or liquids.

They have relatively low melting points and boiling points.

This is because simple molecules such as carbon dioxide and water have weak forces *between* the molecules. Forces of attraction between molecules are known as intermolecular forces. These weak intermolecular forces are easily broken to allow the molecules to move randomly around away from each other.

These substances do not conduct electricity because the molecules do not have an overall electric charge.

① **Explain why carbon dioxide has a boiling point of −78.5 °C.**

Intermolecular forces

Carbon dioxide and water are simple molecules that have strong **covalent bonds** within the molecule. The carbon atom does not break its bonds with the oxygen atoms when carbon dioxide changes state. These are strong covalent bonds formed by electrons sharing.

Melting or boiling involves pulling the molecules away from each other. These substances have only weak forces *between* the molecules (**intermolecular forces**). It is these intermolecular

solid

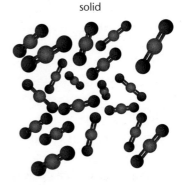

gas

Figure 2.22 CO_2 at low temperatures and CO_2 above boiling point

(a) a model of CO_2

(b) model of H_2O

Figure 2.23 The strong covalent bonds within molecules do not break easily.

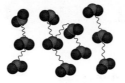
Figure 2.24 Forces *between* molecules have to be overcome for a substance to boil. These are *intermolecular* forces.

forces that are overcome, not the covalent bonds between the atoms, when the substance melts or boils.

Covalent bonds are about 100 times stronger than weak intermolecular forces.

The intermolecular forces increase with the size of the molecules, so larger molecules have higher melting and boiling points.

Substance molecules	CO	CO_2	SO_2	SO_3
Melting point °C	−205	−55.6	−72	17
Boiling point °C	−191.5	−78.5	−10	45

2 Suggest why the boiling point of carbon dioxide, CO_2, is higher than the boiling point of carbon monoxide, CO.

3 Propane, C_3H_8, has a melting point of −188 °C and butane, C_4H_{10}, −140 °C. Suggest why.

4 Nitrogen has a very strong triple covalent bond but a very low boiling point of −195.8 °C. Explain why.

Predicting the properties of small molecules

There are weak *intermolecular* forces *between* the molecules, which affect the physical properties of small molecules.

* As they have weak intermolecular forces between the molecules they are easy to separate so the substances have low melting points.
* As there are no charge carriers available (neither **free electrons** nor free ions) they do not conduct electricity.

However, it is difficult to predict trends in boiling points and melting points as there are other factors involved. Boiling point and melting point not only depend on the *size* of the molecule but on other factors such as its shape. So CH_4 (tetrahedral shape, at −164 °C) has a much lower boiling point than CO_2 (linear, at −78.5 °C) as the molecules cannot get as close to each other.

5 Explain why pure water does not conduct electricity very well, but sea water does.

6 Pentane is a straight chain molecule. 2,2-dimethylpropane has exactly the same molecular formula but is more like a sphere. Explain which is likely to have the higher boiling point.

7 Fluorine, F_2, has a boiling point of −188 °C and chlorine, Cl_2, a boiling point of −34 °C. Suggest why they have different boiling points.

DID YOU KNOW?

Small molecules with strong covalent bonds and weak intermolecular forces make up bulk substances that boil on heating but do not undergo thermal decomposition.

DID YOU KNOW?

Predicting boiling points is not as easy as it seems as there are other forces at work too which depend on the *shape* of the molecule as well.

KEY INFORMATION

The atomic mass of C is 12 and of S is 32.

Polymer structures

Learning objectives:

- identify polymers from diagrams showing their bonding and structure
- explain why some polymers can stretch
- explain why some plastics do not soften on heating.

KEY WORDS
....

intermolecular
force
monomer
polymer
repeating unit

Polymers such as poly(ethene), PVC, PTFE and nylon are used in many areas of everyday life. These polymers are very large molecules. They are made in different ways and have different properties. Can we link their properties to their structure as chemicals?

Polymers

Polymers are very large molecules. The atoms in the polymer molecules are linked to other atoms by strong covalent bonds and so form a long chain.

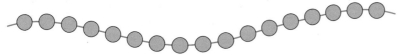

Figure 2.25 Model of polymers such as polyvinyl chloride

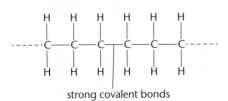

Figure 2.26 The bonds between the atoms in a polymer chain are strong covalent bonds.

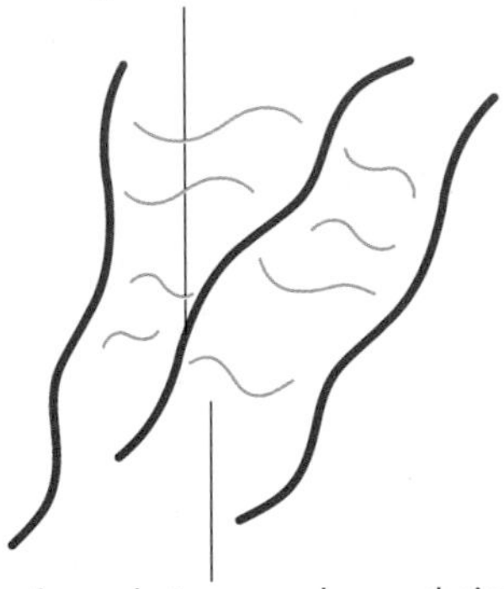

The intermolecular forces *between* polymer molecules are weaker. These intermolecular forces are weak enough to allow the polymer chains to slide other each other but not so weak as to allow the chains to be pulled apart. This means that the polymer can be stretched without breaking.

1 **Poly(ethene) is a solid at room temperature not a liquid. Suggest why.**

Intermolecular forces

Chains of polymers are held together to make a bulk substance.

- The atoms of the **monomers** along the chains in a polymer are held together by strong covalent bonds. These are bonds within the polymer molecules and so are *intra*molecular bonds.
- The chains of the polymer are held together by weak forces of attraction. These are forces between polymer molecules and so are *intermolecular forces*.

strong covalent bonds

weaker forces between polymer chains

Figure 2.27 Strong bonds join atoms into a polymer chain. Weaker forces between the chains hold the chain together.

KEY INFORMATION

Remember *intra*molecular bonds are bonds joining atoms *within* a molecule. *Inter*molecular forces are forces *between* chains of molecules.

Plastics are substances made from polymers such as poly(ethene), polyvinylchloride and nylon. They have weak intermolecular forces of attraction between the polymer molecules so they have low melting points and can be stretched easily as the polymer molecules can slide over one another.

Although the intermolecular forces between polymer molecules are weaker than the strong bonds along the chains, there are many of these forces so together they are relatively strong and so these substances are solids at room temperature. The weaker the force, the lower the melting point.

Type of intermolecular force or bond	Property 1	Property 2
weak intermolecular forces of attraction	stretches easily	lower melting point
strong cross-links – chemical bonds between the chains	rigid	higher melting points

A polymer is a long chain and can be represented by a **'repeating unit'**. In Figure 2.28, the repeating unit has two carbon atoms joined, with four hydrogen atoms; outside the bracket is an 'n' – that means this unit is joined to itself 'n' times into a very long chain.

poly(ethene)

Figure 2.28 Show the repeating unit in brackets. 'n' is a very large number.

2 **Figure 2.28 shows the 'repeating unit' of a polymer chain. Draw three repeating units of this chain.**

3 **Suggest why the polymer melamine cannot stretch and is rigid.**

Bonding and structure

Polymers can be recognised from their structure as they form long chains of repeating units.

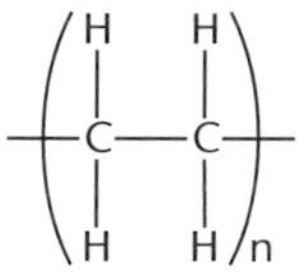

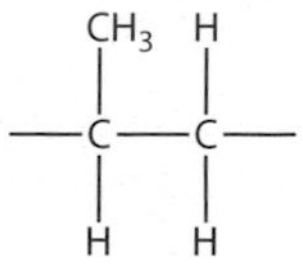

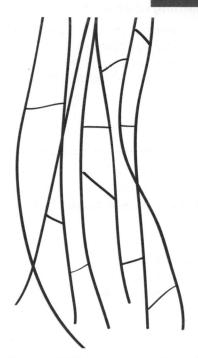

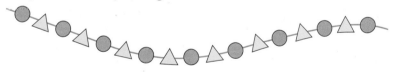

Figure 2.30 Model of polymers such as nylon

Each repeat unit is made up from a group of atoms bonded together with strong covalent bonds. The group is shown inside a pair of brackets with two bonds through the brackets, showing each bond that joins to the next repeating units. As there are many repeating units, the value 'n' is written outside to represent a large number.

4 **Draw three repeating units of the polymer made from the unit:**

Figure 2.29 Polymer chains with weak intermolecular forces (as in Figure 2.27) can slide over one another, chains with cross-links cannot slide.

KEY INFORMATION

Remember the two brackets have a bond through each one. This signifies that the bonds are continuing to join other repeat units.

DID YOU KNOW?

Gore-Tex is a registered product made by expanding PTFE.

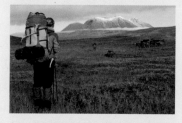

Figure 2.31 Gore-Tex is a popular polymer used by walkers.

Giant covalent structures

KEY WORDS

diamond
giant covalent
    structure
graphite
silicon dioxide

Learning objectives:

- recognise giant covalent structures from bonding and structure diagrams
- explain the properties of giant covalent structures
- recognise the differences in different forms of carbon.

Amethyst, pink quartz and diamond are gems that have giant structures, with bonds in all directions, which make them hard.

Quartz is a giant covalent structure made of **silicon dioxide**. It is the second most abundant mineral in the Earth's crust. It is the biggest component of sand and can also be coloured with other substances, as in some gems. **Diamonds** are not abundant and are giant structures made of carbon.

Structure and bonding

Substances that consist of **giant covalent structures** are solids. All of the atoms in these structures are linked to other atoms by strong covalent bonds. It takes a lot of energy to overcome these strong covalent bonds to melt or boil these substances. These substances with giant covalent structures have very high melting points.

Examples of giant covalent structures are:

- diamond (a form of carbon)
- silicon dioxide (silica or quartz).

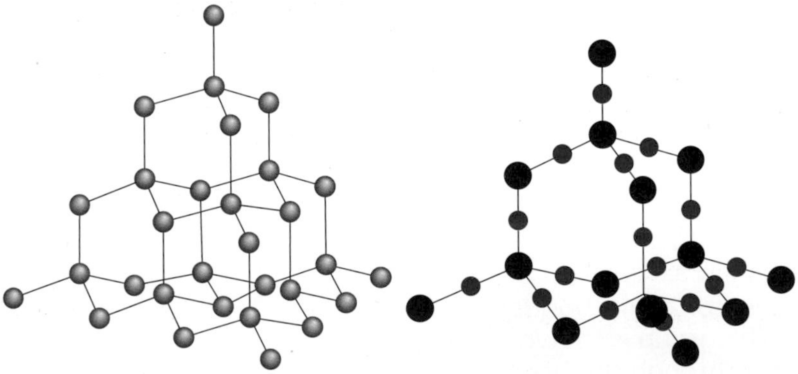

Figure 2.32 Pure quartz is silicon dioxide which is colourless. It can be coloured naturally with iron or manganese to make amethyst.

Figure 2.33 Diamond and silicon dioxide both have giant covalent structures.

KEY INFORMATION

Remember you should aim to be able to recognise the different structures of carbon.

1. **Describe one difference and one similarity in the giant covalent structures of diamond and silicon dioxide.**

2. **Describe the difference between a molecule like water, H_2O, and a giant covalent structure like silicon dioxide.**

The table shows their physical properties.

Diamond	Silicon dioxide
lustrous (shiny), transparent and colourless	white crystalline solid
very hard	very hard
very high melting point	very high melting point
insoluble in water	insoluble in water
does not conduct electricity	does not conduct electricity

3 **Explain why the properties of silicon dioxide are quite similar to the properties of diamond.**

4 **Which two properties of diamond make it suitable for use in high temperature drilling tools?**

Three-dimensional structures

Diamond is not the only giant covalent structure that carbon forms. It also forms **graphite**. The structure of graphite is hexagonal rings of carbon atoms in layers. The bonds do not act in all directions.

Covalent bonding usually means that all the available electrons are shared, so covalent compounds do not usually conduct electricity. Graphite is an exception. Graphite does conduct electricity as it has delocalised electrons that can move through the layers of its structure.

Large structures such as quartz and diamond are hard because of a regular three-dimensional pattern in the bonding of their atoms.

The melting point depends on the energy required to break all the bonds joining the atoms or molecules. As the giant covalent structure has many strong covalent bonds between its atoms in many directions, the energy needed will be high so the melting point will be high.

Covalent compounds are generally insoluble.

5 **Explain why graphite is not as hard as silicon dioxide.**

6 **Suggest why graphite bricks are suitable for use in furnaces.**

Diamond and graphite

Diamond and graphite are both made of carbon atoms, but their structures are very different.

7 **Suggest why graphite can be used in high temperature electrodes.**

8 **Suggest why diamond and silicon dioxide crystals do not conduct electricity.**

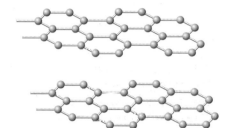

Figure 2.34 Graphite has a giant covalent structure, but it does conduct electricity.

DID YOU KNOW?

Diamond and graphite are both made of carbon atoms, but their structures are very different.

DID YOU KNOW?

Graphite has properties that work in one direction but not at right angles.

Properties of metals and alloys

KEY WORDS
....
alloy
distort
ductile
malleable

Learning objectives:

- identify metal elements and metal alloys
- describe the purpose of a lead-tin alloy
- explain why alloys are harder than pure metals due to the distortion of the layers of atoms.

Making alloys has been carried out for centuries. That early human history is divided into the Stone Age, the Bronze Age and then the Iron Age shows how important bronze was. Bronze was the first metal alloy – copper and tin were both extracted from their ores in rocks thousands of years ago, and later the mixture of these metals was used to make bronze.

Figure 2.35 A bronze shield from the Bronze Age of the Greek Empire. Objects, tools and weapons were made from bronze as it is harder than either tin or copper.

Metals in alloys

Pure copper, gold, iron and aluminium are too soft for many uses. These metals are mixed with other metals to make **alloys**. Most metals we use every day are alloys.

An alloy is a mixture of a metal element with another element.

Alloy	bronze	brass	steel
Elements in alloy	copper and tin	copper and zinc	iron and carbon

Alloys have different properties from the metals that are in them.

Most metals have high melting and boiling points.

Sometimes a metal is needed that has a lower melting point that can then solidify to make a join. This metal alloy is called solder.

Solders are made of lead and tin. Solder has a melting point that is lower than either lead or tin.

Metal	lead	tin	solder
Melting point °C	327	232	183

1 What is the difference between the composition of the alloys brass and bronze?

2 Suggest some advantages of using the alloy steel rather than iron.

3 'Alloys never contain non-metals.' Show whether this statement is true.

Giant structures

In pure metals, atoms are arranged in layers. Pure metals are too soft for many uses and so are mixed with other metals to make alloys which are harder.

Figure 2.36 Solder is an alloy with a lower melting point than tin or lead. If melted it can be used to make or mend metal objects or wiring.

Metals have giant structures of atoms with strong metallic bonding between the positive metal ions and the delocalised electrons. Overcoming this bonding takes a great deal of energy so this means that most metals have high melting and boiling points.

Through this giant structure of metal ions moves the delocalised electrons which gives the metal the ability to conduct electricity. An electrical current is the movement of delocalised electrons through the lattice of ions.

In the giant structures of metals, the layers of atoms are able to slide over each other. This means metals can be bent and shaped. They can be hammered flat (metals are **malleable**) and they can be pulled into a wire (metals are **ductile**).

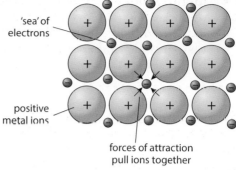

'sea' of electrons

positive metal ions

forces of attraction pull ions together

Figure 2.37

(a) 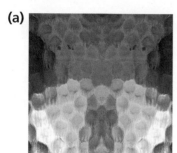   **(b)**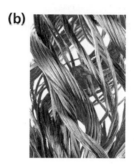

Figure 2.38 Copper is (a) malleable and (b) ductile.

The different sizes of atoms in an alloy **distort** the layers in the structure, making it more difficult for them to slide over each other, so alloys are harder than pure metals.

4 Explain, in terms of the packed atom structure, why copper metal can be extruded into wires.

5 Explain why copper can conduct electricity.

6 Sterling silver contains 92.5% silver, the remainder being copper.

   a Explain whether pure silver or sterling silver is more malleable.

   b Calculate the mass of copper in 1.0 g of sterling silver.

Aluminium

Aluminium is a very useful metal for use in aircraft as it has such a low density and (once it forms a layer of oxide) it is resistant to corrosion. However, it is too soft if it is not alloyed with other metals. The other metals used can be copper, magnesium, manganese and tin. For example, a aluminium and copper alloy is called Duralumin.

7 Explain, in terms of atoms, why alloys are harder and more useable than the main metal within them.

8 Explain why adding copper to aluminium makes an alloy which is harder than both the constituent metals.

9 Copper is a very good electrical conductor. Suggest why alloying copper decreases its conductivity.

DID YOU KNOW?

Smart alloys are alloys that can return to their original shape if heated to a different temperature. This is called a 'shape memory'. Bendable glasses can be made from frames consisting of smart alloys. One smart alloy is 'nitinol'. It is made from nickel and titanium.

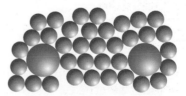

Figure 2.39 Layers of alloy atoms, which are distorted layers due to the different sizes of the atoms.

REMEMBER!

Think about the different sizes of atoms.

Diamond

KEY WORDS

diamond
tetrahedral
lattice

Learning objectives:

- identify why diamonds are so hard
- explain how the properties relate to the bonding structure of diamond
- explain why diamond differs from graphite.

Diamonds **are made from carbon and are the hardest natural substance known. This makes them ideal for industrial cutting tools and grinding wheels. When diamonds form they have cleaving planes that allow them to be shaped and cut to give reflective surfaces that make them one of the most beautiful jewels.**

Structure of diamonds

Diamonds are made of carbon atoms. Each carbon atom forms four bonds. The four bonds form as far from each other as possible. They form in the shape of a tetrahedron.

In diamond, each carbon atom forms four covalent bonds with other carbon atoms in all directions.

A giant covalent structure is made.

Figure 2.40 Diamonds are beautiful but also useful.

Figure 2.41 A carbon atom model showing four bonds in the shape of a tetrahedron

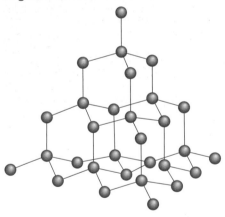

Figure 2.42 Diamond is a giant covalent structure with strong covalent bonds in all directions.

Diamond is very hard, has a very high melting point and does not conduct electricity. There are many strong covalent bonds in diamond in different directions in a 3D structure, which makes diamond very hard.

1. **Describe how a carbon atom bonds.**

2. **Explain why diamonds are so hard.**

3. **Both silicon dioxide, SiO$_2$, and diamond are giant covalent structures. Suggest why silicon dioxide has a lower melting point than diamond.**

Covalent structures

Each carbon atom in diamond makes four covalent bonds with other carbon atoms in all directions in a giant covalent structure, so diamond is very hard. As all the bonds need to break for it to melt, this requires a great deal of energy. Diamond has a very high melting point. All the outer electrons of carbon are used so it does **not conduct electricity**.

4 Explain the difference in electrical properties between diamond and graphite.

5 Explain why each carbon atom in diamond can form four covalent bonds.

Atomic structure and covalent bonding

Carbon atoms have four electrons in the outer electron shell.

In diamond every carbon atom is joined to four others in a three-dimensional **tetrahedral lattice**. The atoms join together by strong covalent bonds that involve electron sharing. This arrangement gives strength in all directions, and requires a large amount of energy to break the bonds, giving a very high melting point of 3350 °C at high pressure.

No delocalised electrons exist in the structure as all outer shell electrons are used in covalent bonds, so diamond does not conduct electricity.

Diamonds are useful as cutting tools and are so rare and expensive that techniques were developed to create synthetic diamonds, known as industrial diamonds. These are not as optically perfect as natural cut diamonds but have the same useful properties and are used in many industries as diamond is the hardest substance.

Figure 2.43 The properties of industrial diamonds make them ideal as cutting tools if not as jewels.

6 Explain how the properties of diamond make it so useful as an abrasive for grinding other materials.

DID YOU KNOW?

Diamond does, however, conduct thermal energy. It does this by transmitting the energy down its covalent bonds as it is in a rigid lattice.

HINTS & TIPS

Carbon atoms in diamond make lots of strong covalent bonds in all directions which is why diamonds are hard.

Graphite

Learning objectives:

- describe the structure and bonding of graphite
- explain the properties of graphite
- explain the similarity to metals.

KEY WORDS

delocalised
  electrons
graphite
hexagonal rings
weak forces

Graphite is a carbon structure that is black and slippery and sticks to paper. It makes good 'lead' for pencils.

The slipperiness also makes graphite a good lubricant, even though it is a solid. Powdered graphite is often used to lubricate door locks and is also used in furnaces and as brake linings. It is a good electrical conductor so it is used in batteries and as powder in carbon microphones.

Graphite layers

Like diamond, **graphite** is also made from carbon atoms. In graphite the carbon atoms are arranged differently. The carbon atoms only make three covalent bonds with other carbon atoms.

The carbon atoms bond to make hexagonal rings which form layers.

Graphite can be easily cut across the layers but not through the layers.

1. State the number of bonds that carbon makes in (a) diamond and (b) graphite.

2. Graphite can be used as a lubricant. State what this tells you about the forces between layers.

Structure and properties

In graphite, each carbon atom forms three strong covalent bonds with three other carbon atoms, forming layers of **hexagonal rings** which have no covalent bonds between the layers.

As there are so many of these strong covalent bonds, which all take a lot of energy to break, graphite has a high melting point.

Only **weak forces** hold the layers together so the layers are free to slide over each other. There are no covalent bonds *between* the layers and so graphite is soft and slippery.

In graphite, only *three* electrons from each carbon atom form strong covalent bonds with electrons from other carbon atoms. The remaining one electron of each carbon atom is *delocalised*. These **delocalised electrons** allow graphite to conduct electricity.

The delocalised electrons move easily along the layers. This is similar to the way that delocalised electrons move in metals. This is why both graphite and metals can conduct electricity. Diamond has no delocalised electrons so cannot conduct electricity.

DID YOU KNOW?

Graphite pencils were first made in 1565. All the graphite came from a mine near the Honister Pass in Cumbria. Clay was added to make different grades of pencil hardness. The graphite in pencils was called 'lead' because the Roman writing implement, the stylus (used for writing on wax tablets), was made of the metal lead.

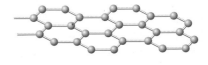

Figure 2.44 Graphite is slippery and can leave marks on paper.

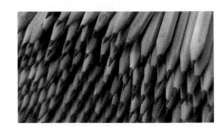

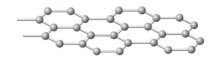

Figure 2.45 Graphite is made of hexagonal rings that stack as layers.

Graphite is also a good conductor of thermal energy, like metals, due to its delocalised electrons too.

3 Compare the organisation of electrons in diamond and graphite.

4 Suggest why graphite is a good thermal conductor.

Uses related to structure

Graphite has a layered arrangement in which each carbon atom is covalently bonded to three others, forming single layers of regular hexagons.

This formation means each carbon atom has an unshared electron in its outer shell, free to move anywhere along the layer. So graphite is an electrical conductor and can be used as electrodes in solutions.

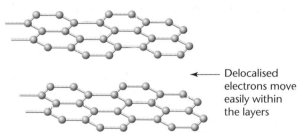

Delocalised electrons move easily within the layers

Figure 2.46 Delocalised electrons can move along layers so graphite is able to conduct electricity.

Graphite has a similar melting point to that of diamond so it is stable at high temperature. Due to these two properties graphite can also be used as electrodes in the electrolysis of *molten* electrolytes.

Figure 2.47 Graphite used as electrodes in a molten electrolyte

The distance between the layers is greater than between bonded atoms and there are no bonding electrons, so the layers in graphite are only weakly attracted to each other. This means that when a force is applied, the layers can slide over each other, which means that graphite can act as a lubricant.

Due to its high melting point it can also be used as a high-temperature lubricant.

Graphite is also a good conductor of thermal energy, like metals, due to its delocalised electrons. However, there is another mechanism for conducting thermal energy, which uses phonons caused by oscillating atoms in a crystalline structure. Diamond has no delocalised electrons but is a better thermal conductor than graphite due to its strong bonds in the 3D lattice structure.

KEY INFORMATION

The weak forces between the layers of hexagonal rings break and allow the layers to slip over one another.

5 Explain (a) why graphite is similar to a metal and (b) why graphite is classified as a non-metal.

6 Explain why graphite can be used as electrodes in molten electrolytes.

7 Explain why graphite has a similar melting point to diamond.

Graphene and fullerenes

Learning objectives:

- explain the properties of graphene in terms of its structure and bonding
- recognise graphene and fullerenes from their bonding and structure
- describe the uses of fullerenes, including carbon nanotubes.

KEY WORDS

cylindrical
fullerene
graphene
nanotube

Carbon does not only exist in the structures of diamond and graphite but also as ball-shaped or cylindrical-shaped structures called fullerenes. The structure of the first one to be discovered looked like a football and was called Buckminsterfullerene and other smaller ones are called 'buckyballs'. The tiny cylindrical tubes are fullerene nanotubes.

Graphene and fullerenes

Graphene is a single layer of graphite and so is one atom thick. It has properties that make it useful in electronics and composites.

It is made up of hexagonal rings of carbon atoms connected to one another by strong covalent bonds.

Carbon atoms in rings can also form hollow 3D shapes. The first one found had 60 carbon atoms. It had rings of six carbon atoms and rings of five carbon atoms. It was named Buckminsterfullerene. Structures of this type are known as **fullerenes**.

Fullerenes are molecules of carbon atoms with hollow shapes. The structure of fullerenes is based on hexagonal rings of carbon atoms but they may also contain rings with five or seven carbon atoms.

Buckminsterfullerene (C_{60}) has a spherical shape, other fullerenes can be in the shapes of spheres or tubes.

There are already a number of medical uses for them but their potential use is still being researched.

1 Compare the structures of graphene and fullerenes.

Carbon nanotubes

Fullerenes can be used for drug delivery into the body, as lubricants, and as catalysts. They can act as hollow cages to trap other molecules. This is how they can carry drug molecules around the body and deliver them to where they are needed, and trap dangerous substances in the body and remove them.

Carbon can also be used to make very small structures called **nanotubes**. Carbon nanotubes are **cylindrical** fullerenes. They have very high length to diameter ratios. Their properties make them useful for nanotechnology, electronics and materials. Some of their special properties are:

- high tensile strength

Figure 2.48 Graphene – hexagonal rings of carbon one atom thick

Figure 2.49 A traditional football can be used as a molecular model for C_{60}.

DID YOU KNOW?

C_{60} was named after the architect Buckminster Fuller as it looked like the geodesic domes that he created.

- high electrical conductivity
- high thermal conductivity.

They are useful:

- as semi-conductors in electrical circuits
- as catalysts
- for reinforcing materials, such as in tennis rackets.

Their potential has not yet been fully developed and new uses are being explored all the time.

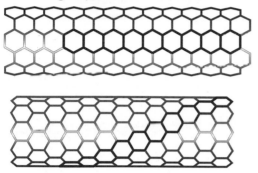

Figure 2.50 Carbon nanotubes. How many carbon atoms are in each ring?

Nanotubes can be used in catalyst systems because atoms of the catalyst can be attached to the nanotubes. The nanotube has a large surface area, so there is more chance that the reactants will collide with the catalyst.

2 Compare the structures of diamond, graphite and a fullerene and explain their differences in bonding.

3 Tensile strength is a measure of how much pulling force it takes to break a material. Suggest why nanotubes have a very high tensile strength.

Properties of graphene

Graphene is a two-dimensional compound as it is only one atom thick. Graphene was isolated in 2003 as a monolayer, and is thermally stable.

Graphene is an electrical conductor through a way known as 'ballistic transport'.

Graphene is the strongest material ever found, stronger than steel and Kevlar.

It is not only strong but elastic too and it can absorb white light. The potential uses of graphene are only just beginning to be researched and this is a major new area in materials science.

4 Predict whether Buckminsterfullerene has a lower or higher melting point than diamond and use your knowledge of their structures to explain reasons for your prediction.

5 Compare graphene and graphite.

6 Suggest why graphene is so strong, unlike graphite.

KEY INFORMATION

Nanoscience refers to structures that are 1–100 nm in size, of the order of a few hundred atoms. They have many applications in medicine in electronics, in cosmetics and sun creams, as deodorants, and as catalysts. New applications for nanoparticulate materials are an important area of research.

You will need to consider advantages and disadvantages of the applications of these nanoparticulate materials, evaluate the use of nanoparticles for a specified purpose and explain that there are possible risks associated with the use of nanoparticles.

KEY CONCEPT

Sizes of particles and orders of magnitude

Learning objectives:

- identify the scale of measurements of length
- explain the conversion of small lengths to metres
- explain the relative sizes of electrons, nuclei and atoms.

KEY WORDS

magnitude
diameter
radius
nanometre

Let's start with the particles we can see. A grain of sand and a grain of sugar are about the same size and are made of crystals. These crystals are made up of much smaller sections that we cannot see.

Orders of magnitude

Placing a tennis ball, golf ball, basketball and table tennis ball in order of size is easy.

unit	basketball	tennis ball	golf ball	table tennis ball
cm	25.0	6.8	4.1	4
m	0.25	0.068	0.041	0.04

Figure 2.51 It is easy to put these in order of diameter.

We can measure objects smaller than these in millimetres.

$1\,m = 1000\,mm$ $1\,mm = 0.001\,m$ or $1\,mm = 10^{-3}\,m$

We can even see objects in the next set of smaller units, the *micrometre*. We measure the width of a human hair in this unit.

$1\,m = 1\,000\,000\,\mu m$ $1\,\mu m = 0.000001\,m$ or $1\,\mu m = 10^{-6}\,m$

After that we need instruments to help us see and measure lengths. We have discussed carbon nanotubes and graphene as a monolayer of carbon atoms. Later, we will discuss large molecules such as DNA. These next sets of objects are in the 'nano-scale.' The unit is the **nanometre**.

$1\,m\ = 1\,000\,000\,000\,nm$
$1\,nm = 0.000000001\,m$ or $1\,nm = 10^{-9}\,m$

1. **Calculate the number of basketballs it would take to make a kilometre.**

2. **A carbon nanotube has a length of 2×10^{-9} m. Calculate the number of nanotubes that would fit in 1 mm.**

Atoms and ions

Going one step further down into the atomic scale:

- the radius of an atom is measured in *picometres* (pm), 10^{-12} m

MATHS

$1\,cm = 0.01\,m$ is the long way to write the conversion.

$1\,cm = 10^{-2}\,m$ is the conversion into *standard form*.

- the radius of a nucleus measured in *femtometres* (fm), 10^{-15} m.

Why is the radius of a nucleus so much smaller than the radius of an atom?

Between the nucleus and the electrons of the atom there is mostly empty space, so neutrons and protons have radii measured in femtometres (fm).

One step even further down, electrons, which are smaller than protons and neutrons, are measured in *attometres* (am) 10^{-18} m.

3 The hydrogen atom has a radius of 2.5×10^{-11} m and its nucleus a radius of 1.75×10^{-15} m. Calculate how many times larger the atom is compared with the nucleus.

4 What is the standard form of a radius 0.000000000001 m?

Radius at the atomic level

The radius of atoms is measured in picometres. Each element has a different atomic radius.

When atoms lose or gain outer electrons for bonding they become ions. Figure 2.52 is a representation of the atomic and ionic radii.

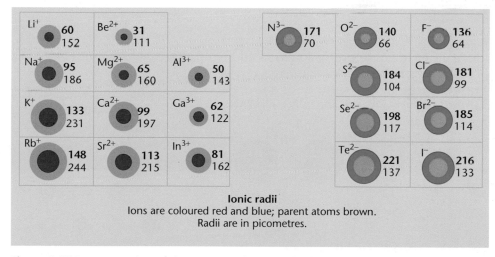

Ionic radii
Ions are coloured red and blue; parent atoms brown.
Radii are in picometres.

Figure 2.52 Representation of the atomic and ionic radii of elements

When atoms join together they make molecules. The smallest molecule is H_2, made of two hydrogen atoms. It has a radius of 0.5 Å whereas Cl_2 has radius of 1 Å. This is 0.1 nm or 100 pm (10^{-10} m).

5 Explain why lithium has an atomic radius of 152 pm but an ionic radius of only 60 pm.

6 Explain why fluorine has an atomic radius of only 64 pm but an ionic radius of 136 pm.

7 Determine which is the greater relative increase in **atomic** radii, Li to Rb or Be to Sr.

MATHS SKILLS

Visualise and represent 2D and 3D shapes

Learning objectives:

- use 2D diagrams and 3D models to:
  - › represent atoms, molecules and ionic structures
  - › represent giant covalent structures
  - › calculate empirical formulae of ionic structures.

KEY WORDS

two dimensional
   representation

hexagonal

model

Even though we can see the beauty of a well-cut diamond it is so hard for us to imagine the individual atoms that are held together in such a rigid structure. It is even harder to imagine that those same atoms, if held in another way, would form the inside of our pencils. Can you use 3D models to represent molecules or giant structures and then draw them?

Models of atoms, molecules and ionic compounds

Atoms are so tiny it is difficult to imagine them, so we need a **model** to help us represent them. We use a circle for a **two-dimensional** (2D) **representation** or a sphere for a three-dimensional representation (3D). This is not *actually* how they look, but is a *representational model*.

DID YOU KNOW?

Atoms are not circles or spheres as they have no distinct outer boundary. Outside the small nucleus of the atom is a 'volume of probability' of finding an electron – you never know where they are!

REMEMBER!

Atoms make covalent bonds by sharing electrons and ionic bonds by transferring electrons.

Figure 2.53 Diagrams of atoms of hydrogen, carbon and oxygen

Figure 2.54 3D models of atoms of hydrogen, carbon and oxygen

If atoms join together with covalent bonds they make small molecules. Examples of these are molecules of hydrogen, oxygen, water and methane.

Figure 2.55 Diagrams of molecules of hydrogen, oxygen, water and methane

Figure 2.56 3D models of atoms of hydrogen, water and methane

If atoms make ions they join together to make ionic compounds. These are often giant ionic lattices.

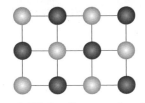

Figure 2.57 2D diagram of sodium chloride lattice

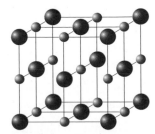

Figure 2.58 3D model of NaCl lattice

1. Draw a 2D diagram of a molecule of carbon monoxide, CO.

2. What do these diagrams or models represent?:

a   b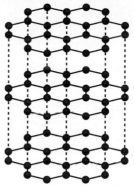

Models of larger structures

Sometimes we need to represent two different structures using the same type of atom, for example, carbon atoms can form graphene, graphite and diamond.

Graphene is made of **hexagonal** rings of carbon atoms in a single layer.

Graphite is made of layers of hexagonal rings held together by weak bonds. A 2D representation of a 3D structure is now needed.

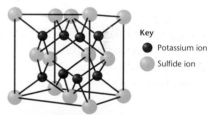
Figure 2.59 Diagram of graphite

Diamond is made of carbon atoms each with 4 bonds in different directions (in a tetrahedral shape). Again a 2D representation of a 3D structure is needed.

Figure 2.60 Diagram of diamond

These structures are giant covalent lattices and not molecules. Buckminsterfullerene (C_{60}) is also made of carbon atoms.

3. Draw a 2D representation of a graphene layer.

4. Suggest why Buckminsterfullerene is called a molecule but graphite is a giant covalent lattice.

Representing giant ionic lattices

To see how many ions are involved in bonding in a crystal it is necessary to have a 3D model or a 2D diagram of a 3D representation of a crystal, as only part of each ion is involved in each section of the large crystal.

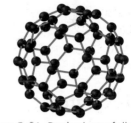
Figure 2.61 Buckminsterfullerene

Figure 2.62 is a representation of potassium ions bonding with sulfide ions. It shows one cube.

Key
● Potassium ion
○ Sulfide ion

Figure 2.62 Potassium sulfide crystal

You need to imagine that another cube is placed on every face of 'our' cube and also at every corner, so that 'our' cube is completely surrounded. You will then need to work out how many other cubes share the face ions. For example, if a surface ion is shared between 2 cubes then each cube claims $\frac{1}{2}$ the surface ion. You need to do the same with the corner ions that are shared with other cubes. You will then be able to add up all the fractions of ions that are shared and be able to work out how many whole ions are used in each 'cube'.

REMEMBER!

You may find it useful to use 'multilink cubes' to help you with this section.

DID YOU KNOW?

Crystallography is the study of using X-ray diffraction to find out the structure of crystals. There are a number of crystal shapes as they depend on the way that the ions pack together.

5. Determine the number of ions or fractions of ions:
   a in the centre of the cube
   b in the centre of a surface
   c at the corners of the cube

6. Use the information in Question 5 to show:
   a the number of potassium ions in one cube
   b the number of sulfide ions in one cube
   c the empirical formula of one cube.

Check your progress

You should be able to:

describe three main types of bonding → explain how electrons are used in the three types of bonding → explain how bonding and properties are linked

represent an ionic bond with a diagram → draw a dot and cross diagram for ionic compounds → work out the charge on the ions of metals and non-metals from the group number of the element

identify ionic compounds from structures → explain the limitations of diagrams and models → work out the empirical formula of an ionic compound

identify single bonds in molecules and structure → draw dot and cross diagrams for small molecules → deduce molecular formulae from models and diagrams

describe that metals form giant structures → explain how metal ions are held together → explain how metallic bonding is enabled by the delocalisation of electrons

use data to predict the states of substances → explain the changes of state → use state symbols in chemical equations

describe the properties of ionic compounds → relate their melting points to forces between ions → explain when ionic compounds can conduct electricity

identify small molecules from formulae → identify polymers from their unit formula → relate the intermolecular forces to the bulk properties of a substance

recognise giant covalent structures from diagrams → explain the properties of giant covalent structures → explain the strength of covalent bonds

identify metal elements and metal alloys → describe the purpose of a lead–tin alloy → explain why alloys have different properties from elements

explain how the properties relate to the bonding in diamond → explain why diamond differs from graphite → explain the similarity of graphite to metals

describe the structure of graphene → explain the structure and uses of fullerenes

Worked example

1 **An element from Group 1, X, bonds with an element from Group 7, Y.**

a Identify the type of bonding.

metallic   (ionic)   covalent   giant

> The answer ionic is correct.

b Draw the dot and cross diagram for the resulting compound. Use X to represent the metal and Y to represent the non-metal. Show the outer shell only.

> **X** has lost an electron so the 1+ charge is missing. **Y** has gained an electron so the 1– charge is missing. Both charges should be at the top right outside the brackets. Electrons should be kept in pairs in a circle.

2 **Explain why both diamond and silicon dioxide are hard with high melting points.**

They have covalent bonds that act in all directions.

> This answer explains what the bonding is like but needs to be linked to the energy required to break the bonds for melting to happen.

3 **Draw diagrams to show how substances change from solids to liquids.**

The particles move faster when they are heated and break away from the solid structure to move more freely.

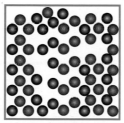

| solid | liquid |

> This answer shows both diagrams and has an added explanation.

4 **Fill in the missing data in the table.**

Substance	Metal	Small molecule	Giant covalent	Ionic
Melting point	High	Low	High	High
Conducts electricity	Yes	No	Yes	Yes when melted, no when solid

> The metal, small molecule and ionic columns are all correct. The student has correctly stated the difference in conductivity between an ionic solid and liquid. Giant covalent structures do not normally conduct electricity. Graphite is an exception and this should be stated.

End of chapter questions

Getting started

1 Identify the type of bonding in potassium bromide, KBr.

    a metallic   **b** ionic   **c** covalent   **d** giant   `1 Mark`

2 Which substance is made of small molecules?

    a carbon dioxide   **b** silicon dioxide   **c** magnesium oxide   **d** sodium sulfide   `1 Mark`

3 Use two 'particle' diagrams to show the differences between a liquid and a gas.   `2 Marks`

4 Substance **R** conducts electricity when solid and is malleable. What is the structure of **R**?

    a simple molecular   **b** giant ionic   **c** giant covalent   **d** giant metallic   `1 Mark`

5 Substance G has a melting point of −33 °C and a boiling point of 52 °C. Explain which state it is in at 20 °C.   `1 Mark`

6 Lithium chloride, LiCl, is made of ions. Explain if it will conduct electricity.   `2 Marks`

7 Match the symbol to the type of particle.   `2 Marks`

K^+	small molecule
NO_2	polymer
-(XZ)-$_n$	ion

Going further

8 Describe the structure of diamond.   `1 Mark`

9 Suggest one structural feature that graphene and fullerenes have in common.   `1 Mark`

10 Cupronickel is an alloy made from copper and nickel. Describe **two** ways in which the properties of the alloy differ from the individual metals.   `2 Marks`

11 Determine the structure of **A** and **B**. Justify your answer.   `4 Marks`

Substance	A	B
Mp and Bp / °C	−205 and −192	420 and 907
Conducts electricity?	No	Yes when solid

12 Metal J has a melting point of 232 °C and metal L has a melting point of 328 °C. When they are mixed together the mixture has a lower melting point than either of these two temperatures. Explain what is being made and why its properties make it useful.   `2 Marks`

More challenging

13 Explain why silver is a good conductor of electricity. `1 Mark`

14 Describe how the electrons are arranged in the bond in chlorine, Cl_2. `1 Mark`

15 The melting point of nylon is higher than that of polythene. Explain why. `2 Marks`

16 Draw the dot and cross diagram for calcium fluoride. Only the outer shell electrons are needed. `2 Marks`

17 Some physical properties for three substances are given in the table below.

Substance	A	B	C
Mp and Bp / °C	1085 and 2562	605 and 1360	3550 and 4830
Conducts electricity?	Yes when solid	Yes when liquid	No
Dissolves in water?	No	Yes	No

   a Identify the structure of **C**. Justify your answer.

   b Explain the difference in the way **A** and **B** conduct electricity. `4 Marks`

Most demanding

18 Methane, CH_4, has low melting and boiling points. Explain why. `2 Marks`

19 Draw a dot and cross diagram for PF_3. Only outer shell electrons need be shown. `2 Marks`

20 Explain the difference between structure and bonding using potassium bromide, KBr, as an example. `2 Marks`

21 Some data on three substances is given in the table.

	Hardness / Mohs' scale	Thermal conductivity / $Wm^{-1}K^{-1}$
Sodium chloride, NaCl	2.5	6.5
Sulfur, S_8	1.5–2.5	0.205
Iron, Fe	4–5	80.4
Silicon dioxide, SiO_2	7	1.4

   a Calculate the relative thermal conductivity of the giant metallic substance compared with the giant ionic substance. Explain the difference.

   b Explain the difference in hardness between the giant covalent substance and the simple molecular substance. `4 Marks`

`Total: 40 Marks`

CHEMICAL QUANTITIES AND CALCULATIONS

IDEAS YOU HAVE MET BEFORE:

CHEMICALS CHANGE DURING REACTION AND ARE NOT DESTROYED

- Iron and sulfur heated together make iron sulfide.
- Magnesium burns in air to make magnesium oxide.
- Wood burns to make ash, smoke and carbon dioxide.

GASES HAVE MASS

- Air in a balloon records a mass on a balance.
- Helium in a balloon has less mass than air in the same balloon.
- Chlorine gas 'rolls' along the ground.

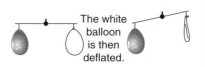

The white balloon is then deflated.

MEASURING QUANTITIES

- Balances measure mass in grams.
- Measuring cylinders measure volume in cm³.
- Stopwatches measure time in minutes and seconds.

RECORDING AND CALCULATING

- Data is recorded and often presented as a graph.
- Units must always be written with data.
- Estimates of the calculation answer should be used as a check.

MAKING SALTS AND CRYSTALS

- When solids are filtered off some solid stays on the paper.
- When solutions are crystallised some crystals stay in the dish.
- Waste of chemicals can be reduced by using careful techniques.

IN THIS CHAPTER YOU WILL FIND OUT ABOUT:

HOW IS MASS CONSERVED IN CHEMICAL REACTIONS?

- The mass of the reactants is the same as the mass of products.
- The same numbers of atoms are on both sides of an equation.
- An equation needs to be balanced by multiples.

WHAT HAPPENS TO MASS CHANGES WHEN A GAS IS GIVEN OFF?

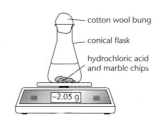

cotton wool bung

conical flask

hydrochloric acid and marble chips

- A gas is often given off during a thermal decomposition reaction.
- Mass seems to decrease in a thermal decomposition reaction.
- A substance reacting with oxygen can seem to increase in mass.

HOW CAN WE MEASURE AMOUNTS OF SUBSTANCES?

- Moles are the chemical measure of amounts of substances.
- Grams and moles interconvert using the molar mass.
- Amounts to be made can be predicted using equations.

HOW CAN WE CALCULATE CONCENTRATIONS OF SOLUTIONS?

- Concentration is the mass of substance in a given volume.
- Concentration is measured in g/dm^3.
- Concentration = mass ÷ volume.

HOW CERTAIN ARE WE ABOUT OUR RESULTS?

- Results are not always certain so we make more measurements.
- We use distributions of results to make estimates of uncertainty.
- We use a range about the mean as a measure of uncertainty.

KEY CONCEPT

Conservation of mass and balanced equations

Learning objectives:

- explain the law of conservation of mass
- explain why a multiplier appears as a subscript in a formula
- explain why a multiplier appears in equations before a formula.

KEY WORDS

balanced symbol equation
conservation of mass
products
reactants

When chemicals react, atoms are not made or destroyed, they are just rearranged. As a result, the total mass is always conserved during a chemical reaction. Because of this, chemical reactions can be represented by balanced symbol equations.

The law of conservation of mass

The law of **conservation of mass** states that 'no atoms are lost or made during a chemical reaction'.

This means that the mass of the **products** equals the mass of the **reactants**.

In a chemical reaction, all the atoms in the reactants are *rearranged* into products. For example, for one type of reaction:

$$AB \ + \ CD \ \rightarrow \ AD \ + \ CB$$

the mass of atoms at the start equals the mass of atoms at the finish. This is consistent with the law of 'conservation of mass'.

1 Identify the missing product to ensure conservation of mass.

$$XY \ + \ ZRT \ \rightarrow \ XR \ + \ ? \ + \ T$$

Balanced equations and formulae

Many compounds contain more than one atom of an element in a formula. The number of atoms of the element is written as a subscript that is written after the element symbol.

For example, these formulae have more than one atom of some of their elements. If there is no number after the element symbol it means 1 atom.

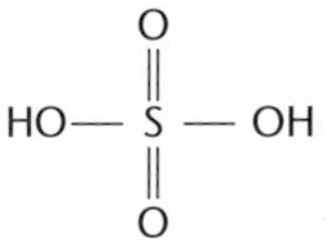

Figure 3.1 Masses of reactants and products are balanced.

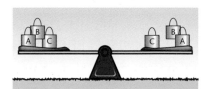

Figure 3.2 Numbers of atoms of the formulae are balanced.

$$HO — \overset{\overset{O}{\|}}{\underset{\underset{O}{\|}}{S}} — OH$$

Figure 3.3 The molecule has 2 H atoms, 1 S atom and 4 O atoms.

KEY SKILLS

Always work out or look up the formula first, then put it into an equation.

carbon dioxide	CO_2	O=C=O	magnesium chloride	$MgCl_2$	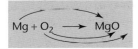
water	H_2O	H—O—H	sodium sulfate	Na_2SO_4	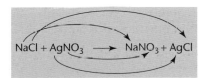
ethane	C_2H_6	H—C—C—H (with H atoms)	calcium nitrate	$Ca(NO_3)_2$	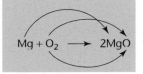

2 Determine the number of hydrogen atoms in ethanoic acid CH_3COOH.

3 How many atoms of each element are in sodium thiosulfate $Na_2S_2O_3$?

4 How many atoms of each element are in aluminium sulfate $Al_2(SO_4)_3$?

Balancing equations

This equation is **balanced**.

$$NaCl + AgNO_3 \rightarrow NaNO_3 + AgCl$$

Here is another equation:

$$Mg + O_2 \rightarrow MgO$$

This equation is *not* balanced.

The two atoms of oxygen will each join with a magnesium atom to make two formula units of MgO.

$$Mg + O_2 \rightarrow 2\,MgO$$

This equation is still not balanced.

If we make two formula units of MgO, there must have been two atoms of Mg as reactants to 'conserve mass'.

$$2Mg + O_2 \rightarrow 2\,MgO$$

The equation is now balanced. There are the same numbers of atoms on the two sides and 'mass is conserved'.

5 Balance the following equations by determining the numerical values for **d, e, f** and **g** in each of them.

  a $\mathbf{d}NaOH + \mathbf{e}H_2SO_4 \rightarrow \mathbf{f}Na_2SO_4 + \mathbf{g}H_2O$

  b $\mathbf{d}C_3H_8 + \mathbf{e}O_2 \rightarrow \mathbf{f}CO_2 + \mathbf{g}H_2O$

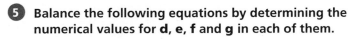

$NaCl + AgNO_3 \longrightarrow NaNO_3 + AgCl$

$Mg + O_2 \longrightarrow MgO$

$Mg + O_2 \longrightarrow 2MgO$

$2Mg + O_2 \longrightarrow 2MgO$

DID YOU KNOW?

You must write the equation of the reaction that actually happened even if it is difficult to balance. You cannot alter the formulae of compounds even if this makes balancing easier.

Relative formula mass

Learning objectives:

- identify the relative atomic mass of an element from the periodic table
- calculate relative formula masses from atomic masses
- verify the law of conservation of mass in a balanced equation.

When looking at the formula of a compound, the relative atomic masses of all the elements can be added to find the mass of the whole compound. The formula mass of the reactants in a chemical reaction must equal the formula mass of the products according to the law of conservation of mass.

Relative atomic mass

Atoms are made up of protons, neutrons and electrons. Protons and neutrons make up the nucleus with the electrons orbiting the nucleus in 'shells'. Most of the atom is empty space with almost all the mass in the nucleus.

A lithium atom has three protons, so has an *atomic number* of 3.

A lithium atom has four neutrons, so has a *mass number* of $3 + 4 = 7$

This can be written as $^{7}_{3}Li$.

Some elements have *isotopes*. Although all the atoms of an element must have the same atomic number (and so have the same number of protons) the atoms do not always have the same atomic mass. They have different numbers of neutrons.

For example, hydrogen has three common isotopes: $^{1}_{1}H$ $^{2}_{1}H$ $^{3}_{1}H$

In a sample of hydrogen, there are some atoms with mass 3, some with mass 2 but most have mass 1. Taken on *average*, the mass of a typical sample of hydrogen is 1.00794. This is not the mean of 1, 2 and 3 but takes into account the proportion of each isotope.

Isotopes are the reason when you look on the periodic table, the atomic masses of some elements are given with numbers after a decimal point. Examples are Cl as 35.5 and Cu as 63.5.

1. **50% of bromine atoms have a mass of 79 and 50% a mass of 81. Work out the atomic mass of bromine.**

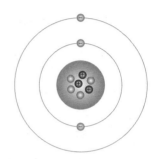

Figure 3.4 The lithium atom has three protons and four neutrons.

DID YOU KNOW?

This accurate value for hydrogen is usually rounded to 1.

The carbon-12 standard

The mass of hydrogen used to be taken as the standard against which to measure the *relative* **atomic masses** (A_r) of other atoms. Later, it was found that better results were obtained if everything was compared with the isotope of carbon that has an atomic mass of 12.

The relative atomic mass of an element is the average mass of an atom of the element compared with the mass of an atom of carbon-12. Values for each element are shown in the periodic table.

2 Show the atomic structure of the three isotopes of carbon, $^{12}_{6}C$, $^{13}_{6}C$ and $^{14}_{6}C$ in terms of protons and neutrons.

3 Explain why relative atomic mass is defined in terms of average mass.

Formula mass

The relative **formula mass** (M_r) of a compound is the sum of the relative atomic masses of the atoms in the numbers shown in the formula.

For example:

M_r of MgO	$= A_r$ of Mg $+ A_r$ of O		
	$= 24$	$+ 16$	$= 40$
M_r Na$_2$O	$= (23 \times 2) + 16$		$= 62$
M_r CO$_2$	$= 12$	$+ (16 \times 2)$	$= 44$
M_r CuCO$_3$	$= 63.5$	$+ 12 + (16 \times 3)$	$= 123.5$
M_r Mg(OH)$_2$	$= 24$	$+ (16 \times 2) + (1 \times 2)$	$= 58$

In a balanced chemical equation, the sum of the relative formula masses of the reactants equals the sum of the relative formula masses of the products.

	ZnCO$_3$	$\rightarrow$	ZnO + CO$_2$
M_r are:	$65 + 12 + (16 \times 3)$		$65 + 16 + 12 + (16 \times 2)$
	$65 + 12 + 48$	$\rightarrow$	$81 + 12 + 32$
	125	$\rightarrow$	125

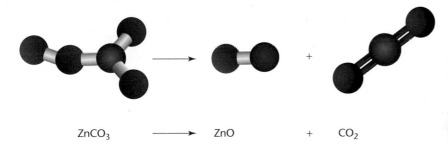

ZnCO$_3$ $\longrightarrow$ ZnO + CO$_2$

4 Work out the formula mass of MgSO$_4$.

5 Calculate the formula mass of Cu(NO$_3$)$_2$.

6 Show that the relative formula masses of reactants and products are equal in this reaction:

$$MgBr_2 + 2AgNO_3 \rightarrow Mg(NO_3)_2 + 2AgBr$$

7 R is a molecule containing carbon and hydrogen. Work out the molecular mass and molecular formula of R.

$$R + 5O_2 \rightarrow 3CO_2 + 4H_2O$$

Mass changes when gases are in reactions

Learning objectives:

- explain any observed changes in mass in a chemical reaction
- identify the mass changes using a balanced symbol equation
- explain these changes in terms of the particle model.

KEY WORDS

gas
mass
particles

We have learned that mass must be conserved in a reaction. The sum of the masses of the reactants must equal the sum of the masses of the products. What is happening when some chemicals seem to lose mass when they are heated?

Losing mass

If baking powder is heated, it gives off carbon dioxide, which makes cakes 'rise'. It seems as if the mass of baking powder is higher before it is heated than after.

Some reactions may seem to involve a change in **mass**. This can usually be explained because a reactant or product is a **gas**. The mass of the gas is often not measured.

A gas can be driven off as a product or taken in as a reactant.

For example: when copper carbonate is heated, it reacts to make copper oxide and carbon dioxide. Where does the carbon dioxide gas go?

Sammy heats 5 g of copper carbonate.

After heating, Sammy measured the mass of the copper oxide.

It had a mass of 3.2 g.

How much gas was made?

The answer is 1.8 g.

How did Sammy work this out?

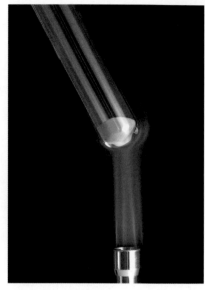

Figure 3.5 Heating copper carbonate

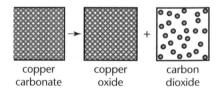

copper carbonate → copper oxide + carbon dioxide

Figure 3.6 Solid copper carbonate when heated turns to solid copper oxide and the gas carbon dioxide.

1. **Jane heats 2 g of zinc carbonate.**

   zinc carbonate → zinc oxide + carbon dioxide

   2 g of zinc carbonate makes 1.3 g of zinc oxide. Explain why the mass of zinc oxide is less than the mass of zinc carbonate.

2. **In the reaction in question 1, what mass of carbon dioxide is produced?**

3. **Copper sulfate crystals change into copper sulfate powder when heated. Water is given off. If 2.5 g of copper sulfate crystals make 1.6 g of powder, how much water is produced?**

KEY INFORMATION

The conservation of mass means that the mass of the products must equal the mass of the reactants.

Gaining mass

Sometimes one of the reactants is a gas.

For example: when a metal reacts with oxygen the mass of the oxide produced is greater than the mass of the metal.

Alesha heats 4 g of copper. When it cools the solid product has a mass of 5 g. Why is there an increase?

The answer is that the copper reacted with 1 g of oxygen from the air to form copper oxide.

Figure 3.7 Blue, hydrated copper sulfate becomes white when heated and the water is driven off.

4 Draw three boxes. Draw the **particle** arrangement for the reactants and products in the reaction of
copper + oxygen → copper oxide

5 When magnesium is heated in air it gains mass. If 2.4 g of magnesium makes 4 g of magnesium oxide, how much oxygen was added from the air?

HIGHER TIER ONLY

Other reactions and limiting factors

If magnesium carbonate is put into acid in a flask and the flask is on top of a digital balance, what is observed? The mass of the flask and its contents decrease.

This is because carbon dioxide is being given off.

$$MgCO_3 + 2HCl \rightarrow MgCl_2 + H_2O + CO_2$$

If the mass decrease is measured every two minutes a graph such as Figure 3.9 could be drawn to use in an investigation.

It is not only carbon dioxide or water vapour that will be produced in a reaction. Other reactions will produce other gases such as hydrogen. These reactions will produce a decrease in mass also.

$$Mg + 2HCl \rightarrow MgCl_2 + H_2$$

If the acid in this reaction is in excess, then the magnesium is the limiting reactant. Once the magnesium has been used up no more hydrogen can be made.

6 Look at the graph in Figure 3.9. Explain when the reaction stopped and why it stopped.

7 The reaction between 8.4 g of $MgCO_3$ and excess hydrochloric acid solution stops after 7 minutes. The mass of the flask and contents has decreased by 4.4 g. Sketch the graph you might obtain when monitoring this reaction.

8 6.54 g of zinc was added to 6.97 g of nitric acid. At the end of the reaction, 2.92 g of zinc remained.

  a Explain which of the reactants was the limiting reactant.

  b State the mass of zinc and mass of nitric acid that reacted.

9 Explain whether mass will be gained or lost in the following reaction. CH_3COOH and H_2O are liquids and O_2 and CO_2 are gases.

$$CH_3COOH + 2O_2 \rightarrow 2CO_2 + 2H_2O$$

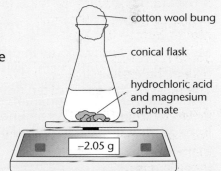

cotton wool bung

conical flask

hydrochloric acid and magnesium carbonate

−2.05 g

Figure 3.8 Measuring the mass of carbon dioxide given off

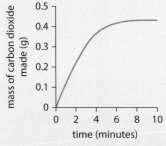

mass of carbon dioxide made (g)

time (minutes)

Figure 3.9 Graph showing mass of CO_2 given off

DID YOU KNOW?

It is possible to work out exactly how much gas should be given off if we know the amount of limiting reactant we start with.

Chemical measurements and uncertainty

KEY WORDS

uncertainty
distribution
  of results
range of
  measurements

Learning objectives:

* understand that all measurements have a degree of uncertainty
* estimate the uncertainty from the distribution of results
* measure uncertainty from the range of a set of measurements and their mean.

Sports reporters may claim a crowd of 15 000 watched a football match. 15 000 is unlikely to be the exact number of spectators, or the true value. The real value is likely to be 15 000 give or take a few. We say there is a 'degree of uncertainty' about the number 15 000 when describing the size of the crowd.

Uncertain measurements

Measuring cylinders come in different sizes. A 50 cm³ cylinder will have a scale with 1 cm³ divisions. You can probably read volumes to half a division, or to 0.5 cm³. However, manufacturers do not guarantee measuring cylinders to be accurate. There is always an **uncertainty** when a measurement is made. This is usually printed on the apparatus.

KEY INFORMATION

Accuracy is how close a measurement or result is to the true value.

KEY INFORMATION

The machines that manufacture measuring apparatus have set tolerance levels. These are the acceptable ranges within which the apparatus measures.

KEY INFORMATION

Uncertainties should have the same number of decimal places as the measurement.

So, 2.2 cm³ ± 0.1 cm³ is correct, but 2.2 cm³ ± 0.15 cm³ is not.

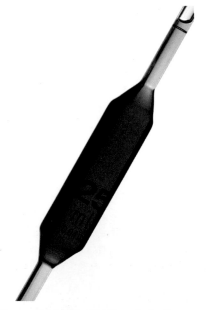

Figure 3.10 This measuring cylinder has an uncertainty of ±0.5 cm³. A volume of 15.0 cm³ should really be written as 15.0 cm³ ± 0.5 cm³. This means that the true measurement will be somewhere between 14.5 cm³ and 15.5 cm³.

Figure 3.11 This 25.00 cm³ pipette has an uncertainty of ±0.04 cm³. Where does the true value lie?

1 A burette used to measure volumes of liquids has an uncertainty of ±0.05 cm³. Emma uses it to measure 25.30 cm³ of water. How much water could Emma actually have measured?

Distribution of results and uncertainty

Two different groups of pupils measured the volume of dilute acid needed to react exactly with 25.0 cm³ of a sodium hydroxide solution. The groups have made a frequency table and histogram to illustrate the **distribution of results**.

Group A – results table

Volume of dilute hydrochloric acid (cm³)	Number of results in this range
12.0–12.9	0
13.0–13.9	3
14.0–14.9	8
15.0–15.9	4
16.0–16.9	0

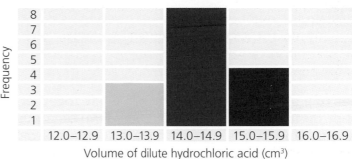

Group B – results table

Volume of dilute hydrochloric acid (cm³)	Number of results in this range
12.0–12.9	3
13.0–13.9	4
14.0–14.9	3
15.0–15.9	1
16.0–16.9	4

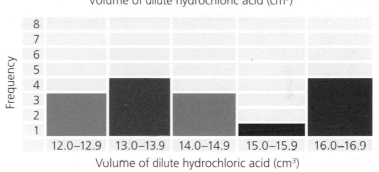

Group A's results are spread between 13.0 cm³ and 15.9 cm³. Group B's are between 12.0 cm³ and 16.9 cm³. Group B's results have a higher uncertainty than Group A's.

2 Which group has the most precise results?

Range of measurements and uncertainty

The **range** of a set of measurements is the difference between the highest and lowest measurements.

range = highest measurement – lowest measurement

These are some class results for a titration experiment:

6.8 g/dm³   7.1 g/dm³   6.9 g/dm³   8.2 g/dm³
6.3 g/dm³   7.4 g/dm³   6.6 g/dm³   5.9 g/dm³

- the range of the results = 8.2 – 5.9 = 2.3 g/dm³

The mean result is found by adding all the results together and dividing by the number of results:

- mean result $= \dfrac{6.8 + 7.1 + 6.9 + 8.2 + 6.3 + 7.4 + 6.6 + 5.9}{8} = 6.9$ g/dm³

The range and mean can be used to calculate the percentage uncertainty.

$$\text{percentage uncertainty} = \frac{\text{range of measurements}}{\text{mean}} \times 100$$

- the percentage uncertainty for the class results is $\dfrac{2.3}{6.9} \times 100 = 33.3\%$

3 What is the difference in meaning between 'range' and 'mean'?

4 A set of measurements has a range of 0.60 cm and a mean of 7.5 cm. What is the percentage uncertainty?

KEY WORDS
....
mole
chemical amount
relative formula mass
Avogadro's constant

objectives:

- describe the measurement of amounts of substances in moles
- calculate the number of moles in a given mass
- calculate the mass of a given number of moles.

Particles of a substance can be packed close together as in copper metal, diamond or ice, or they can be moving rapidly and randomly about as in water vapour or oxygen gas. What is the one measurement which allows us to compare the amount of substance in each chemical?

HIGHER TIER ONLY

Measurement of amounts

Chemical amounts are measured in **moles**. The symbol for the unit mole is mol.

The measurement of amounts in moles can apply to:

atoms	ions	formulae
molecules	electrons	equations

For example:

In one mole of carbon (C) the *number* of atoms is the same as the number of molecules in one mole of carbon dioxide (CO_2).

The mass of one mole of a substance in grams is numerically equal to its **relative formula mass**.

Number of atoms in 1 mole C = number of molecules in 1 mole CO_2	
Carbon has a formula of C	Carbon dioxide has a formula of CO_2
C has a relative atomic mass of 12	CO_2 has a relative molecular mass of $12 + (2 \times 16) = 44$
Mass of 1 mole = 12 g	Mass of 1 mole = 44 g

Figure 3.12 This lump of carbon (C) and jar of carbon dioxide (CO_2) may not look the same but they have the same amount of substance. There is the same number of atoms in the carbon lump as the number of molecules of carbon dioxide in the jar.

1 Work out the mass of one mole of H_2O.

[Relative atomic mass of H is 1, relative atomic mass of O is 16]

2 Work out the mass of three moles of KBr. [Use the periodic table to help you]

Avogadro's constant

Single atoms cannot be measured for mass in an experiment or in usage – they are too small. Instead, they are used in standard amounts.

The mole is a unit for a standard *amount* of a substance. One mole of any substance contains the same number of particles, atoms, molecules or ions as one mole of any other substance.

The number of atoms, molecules or ions in a mole of a given substance is the **Avogadro's constant** .

The value of the Avogadro's constant is 6.02×10^{23} per mole (or mol^{-1})

3 Calculate the number of particles in two moles of helium, He.

DID YOU KNOW?

Avogadro's hypothesis was developed in 1810, over 200 years ago.

Calculating molar mass

The name for the mass of one mole is the molar mass and its unit is g/mol or $gmol^{-1}$.

A_r is the abbreviation used for relative atomic mass. The mass of one mole of an element that consists of atoms is numerically equal to the relative atomic mass, A_r on the periodic table. For example:

- Ne has a relative atomic mass, A_r, of 20.
- 1 mole of neon, Ne has a molar mass of 20 g/mol.

The mass of one mole of a molecule is found by adding up all the relative atomic masses in the formula.

Oxygen gas has the formula O_2.

- The relative atomic mass, A_r, for oxygen is 16.
- In O_2, there are two oxygen atoms.
- The molar mass for O_2 is 16 + 16 = 32 g/mol.

The mass of one mole of a compound is found by adding up all the relative atomic masses in the formula.

- Calcium carbonate has the formula $CaCO_3$.
- Relative atomic masses, A_r, are Ca = 40, C = 12, O = 16.
- The molar mass of $CaCO_3$ is 40 + 12 + (3 × 16) = 100 g/mol.

If the formula contains brackets, these must be considered in the calculation:

- Magnesium nitrate has the formula $Mg(NO_3)_2$.
- Relative atomic masses, A_r, are Mg = 24, N = 14, O = 16.
- Relative formula mass of $Mg(NO_3)_2$ = 24 + (2 × 14) + 2 (3 × 16) = 148.
- The molar mass is 148 g/mol.

4 Work out the molar mass of:

    a nitrogen gas, N_2

    b zinc oxide, ZnO

    c magnesium carbonate, $MgCO_3$

    d ammonium sulfate, $(NH_4)_2SO_4$

5 Work out the number of moles of water in 72 g.

6 $2H_2 + O_2 \rightarrow 2H_2O$

    a Work out the number of moles of O_2 needed to make four moles of water.

    b Calculate the number of grams of H_2 needed to make four moles of water.

KEY INFORMATION

If the formula contains brackets, then the mass can be calculated in two ways, e.g. $Mg(NO_3)_2$ = 24 + (2 × 14) + 2 (3 × 16) = 148 or 24 + 2 [14 + (3 × 16)] = 148

Amounts of substances in equations

Learning objectives:

- calculate the masses of substances in a balanced symbol equation
- calculate the masses of reactants and products from balanced symbol equations
- calculate the mass of a given reactant or product.

We now know that chemists measure the amount of substance as a number of moles. We need to use a balanced equation to see how many moles of reactant will produce the number of moles of product. But how do we know how many grams to measure out? What is the reactant mass that we need to use? How much product mass will we get?

HIGHER TIER ONLY

Masses of substances from equation

The masses of reactants and products can be calculated from balanced symbol equations. For example, looking at the reaction between magnesium and oxygen to form magnesium oxide:

$$2Mg + O_2 \rightarrow 2MgO$$

we can see that:

2 moles of Mg react with 1 mole of O_2 to produce 2 moles of MgO.

We know that the relative formula mass of:

Mg is 24	O_2 is 16×2	MgO is $24 + 16$

So if 2×24 g of Mg reacts with 32 g of O_2 then 2×40 g of MgO is made.

$$2Mg + O_2 \rightarrow 2MgO$$
$$48 \text{ g} + 32 \text{ g} \rightarrow 80 \text{ g}$$

If 4.8 g of Mg is used then 8.0 g of MgO is made.

This can be written as:

$$\frac{\text{Molar mass of substance A}}{\text{Mass of reactant A}} = \frac{\text{Molar mass of substance B}}{\text{Mass of product B}}$$

$$\text{or Mass of B} = \text{Mass of A} \times \frac{\text{Molar mass of B}}{\text{Molar mass of A}}$$

Figure 3.13 Magnesium reacting with oxygen to form magnesium oxide

KEY INFORMATION

Don't forget that as these are *ratios* the calculation can also be rearranged to:

$$\frac{\text{Mass of substance A}}{\text{Molar mass of A}} = \frac{\text{Mass of substance B}}{\text{Molar mass of B}}$$

1 Calculate the product mass of MgO made from 6.0 g of Mg.

2 Calculate the reactant mass of Mg needed to make 2.0 g of MgO

Measuring the number of moles in different ways

Chemical equations can be interpreted in terms of **moles**. Another example is:

$$Mg + 2HCl \rightarrow MgCl_2 + H_2$$

This shows that one mole of magnesium reacts with two moles of hydrochloric acid to produce one mole of magnesium chloride and one mole of hydrogen gas. These are the ratios in which reactants and products relate. The actual amount of moles made will be in these ratios but can be measured in terms of mass, concentration or volume.

3 Determine the number of moles of H_2O that will be made from six moles of propane on combustion with O_2.

$$C_3H_8 + 5O_2 \rightarrow 3CO_2 + 4H_2O$$

Predicting masses

Moles can also be used to predict masses from equations.

This is the equation for burning a fuel called heptane, C_7H_{16}:

$$C_7H_{16} + 11O_2 \rightarrow 7CO_2 + 8H_2O$$

What mass of carbon dioxide, CO_2, is formed when 100 g of C_7H_{16} is burned? Answer:

Stage 1: find the **molar masses**.

Molar mass of C_7H_{16} is $(7 \times 12) + (16 \times 1) = 100$ g

Molar mass of CO_2 is $(1 \times 12) + (2 \times 16) = 44$ g

Stage 2: use the numbers of moles from the equation to find the ratio.

From the equation, 1 mole of heptane produces 7 moles of CO_2.

The mass of 7 moles of CO_2 is $7 \times 44 = 308$ g

So 100 g of heptane will produce 308 g of carbon dioxide.

Use the relative atomic masses in the periodic table to help you answer Questions 4 and 5.

4 Find the product mass of ZnO made when 1.25 g of $ZnCO_3$ are thermally decomposed.

$$ZnCO_3 \rightarrow ZnO + CO_2$$

5 Find the mass of $CuCO_3$ needed to make 7.95 g of CuO.

DID YOU KNOW?

The number of moles of Mg that is actually used can be found from the mass of Mg used measured in grams. The number of moles of HCl can be found from the concentration of the acid in grams/dm³. The number of moles of H_2 made can be found from the volume collected compared with the molar volume.

Using moles to balance equations

Learning objectives:

- convert masses in grams to amounts in moles
- balance an equation given the masses of reactants and products
- change the subject of a mathematical equation.

KEY WORDS

balanced symbol
  equation
moles
thermal decomposition
molar mass

We have already learned how to calculate molar masses and use balanced equations **to calculate the mass of a product that should be made from a given mass of reactant. Can we turn these calculations around so that we can use the mass of product formed to find the number of moles that react and so balance an equation?**

HIGHER TIER ONLY

Finding the number of moles

The balancing numbers in a symbol equation can be calculated from the masses of reactants and products by converting the masses in grams into amounts in moles and converting the numbers of **moles** to simple whole number ratios.

Let's look at the **thermal decomposition** of magnesium carbonate, which produces magnesium oxide (MgO) and carbon dioxide (CO_2). We find that if 42 g of $MgCO_3$ is heated then 20 g of MgO is produced.

We have already learned that:

mass of chemical = **molar mass** × number of moles

So this equation can be rearranged to find the number of moles:

$$\text{number of moles} = \frac{\text{mass of chemical}}{\text{molar mass}}$$

So to calculate the number of moles in 20 g of MgO:

First calculate the molar mass M_r of MgO:

(A_r Mg = 24, A_r O = 16) M_r MgO = 24 + 16 = 40 g/mol.

Then use the equation: $\text{number of moles} = \dfrac{\text{mass of chemical}}{\text{molar mass}}$

$$\text{number of moles} = \frac{20}{40} = 0.5 \text{ moles}$$

If we heated 42 g of $MgCO_3$, then we would get 20 g of MgO.

Figure 3.14 The thermal decomposition of $MgCO_3$.

KEY INFORMATION

In this case 22 g of CO_2 would have been driven off: remember the conservation of mass.

How many moles is 42 g of $MgCO_3$?

Using the same calculation as above we would calculate:

First the molar mass M_r of $MgCO_3$: (A_r Mg = 24, A_r C = 12, A_r O = 16)

$$M_r\ MgCO_3 = 24 + 12 + (3 \times 16) = 84\ g/mol.$$

Then use the equation: number of moles $= \dfrac{\text{mass of chemical}}{\text{molar mass}}$

number of moles $= \dfrac{42}{84} = 0.5$ moles

1 **What is the mass of MgO that would be produced by 84 tonnes of $MgCO_3$?**

Balancing the equation

$$MgCO_3 \rightarrow MgO$$

So this calculation tells us that:

0.5 moles $MgCO_3$ → 0.5 moles MgO

so thus 1 mole of $MgCO_3$ makes 1 mole of MgO.

The ratio is 1:1

The ratio of moles in the equation is therefore:

$$MgCO_3 \rightarrow MgO + ?\ CO_2$$

But what about the number of moles of CO_2? Looking at the formula there must be 1 mole of CO_2 produced.

We need to check by mass. We saw that to conserve mass 22 g of CO_2 would have been produced. Molar mass of CO_2 is 44 g/mol.

number of moles $= \dfrac{\text{mass of chemical}}{\text{molar mass}} = \dfrac{22}{44}$

So again 0.5 moles of $MgCO_3$ makes 0.5 moles of MgO and 0.5 moles of CO_2, which is a 1:1:1 ratio.

The **balanced equation** is therefore:

$$MgCO_3 \rightarrow MgO + CO_2$$

Applying the calculations

Aluminium oxide, Al_2O_3 produces aluminium, Al and oxygen, O_2

If 204 g of Al_2O_3 produce 108 g of Al work out the number of moles of Al_2O_3, Al and O_2 involved and hence write out the full balanced equation.

2 **Explain how you worked out the number of moles of Al_2O_3 and Al from the masses given.**

3 **Explain how you worked out the number of moles of O_2 and how you deduced the mole ratio of the three substances.**

3.7

DID YOU KNOW?

In this example we found that 0.5 moles of $MgCO_3$ produced 0.5 moles of MgO. So we can say that 1 mole of $MgCO_3$ would produce 1 mole of MgO.

As these are ratios we can also say 17 tonnes of $MgCO_3$ would produce 17 tonnes of MgO.

Concentration of solutions

Learning objectives:

- relate mass, volume and concentration
- calculate the mass of solute in solution
- relate concentration in mol/dm^3 to mass and volume.

We have learned before that a solution forms when a solute is dissolved in a solvent and that solutions can be dilute or concentrated. Dilutions are often critical, such as making the correct formulations of medicines or baby milk formula. How do we make sure we have the correct concentration?

Checking the correct units

Many chemical reactions take place in **solutions** and often the **concentration** needs to be known. If a volume of **solvent** is used, say 100 cm^3, the number of particles of **solute** in the solvent is lower if the solution is more dilute and is higher if the solution is more concentrated.

Figure 3.15 The concentration increases as the number of solute particles in a fixed volume increases.

If a 100 cm³ solution contains 2.5 g of solute X, then the concentration is 2.5 g/100 cm³.

This is because the *concentration* of a solution can be measured as the *mass of solute dissolved per specified volume of solution*

or $$\text{concentration} = \frac{\text{mass of solute}}{\text{volume}}$$

The standard unit of concentration used in the laboratory is: grams per dm³ so the units are then g/dm³.

1 dm³ is 1000 cm³ so the concentration of X is 25 g/dm³

1 If the next, fourth, solution in Figure 3.15 was more concentrated, describe how you would draw the solution in the beaker.

2 Draw a diagram of the middle beaker in Figure 3.15 if the solution was the same concentration but the volume was only 50 cm³.

Calculating from concentrations

A solution has a concentration of 6 g/dm³. What is the mass of solute Z that was added to 100 cm³ to make this solution?

$$\text{Concentration} = \frac{\text{mass}}{\text{volume}}$$

$$\text{Concentration} = \frac{6}{1000} \text{ g/dm}^3 \quad \text{and Concentration} = x \text{ g/100 cm}^3$$

Therefore: $\dfrac{x}{100} = \dfrac{6}{1000}$ so $x = \dfrac{6 \times 100}{1000} = 0.6$ g

3 Put the following solutions into order with the most dilute first.
a 20 g/100 cm³   b 20 g/1000 cm³   c 8 g/50 cm³

4 Calculate the concentration of the following solutions in g/dm³.
a 3.2 g in 100 cm³   b 3.2 g in 250 cm³   c 6.4 g in 500 cm³

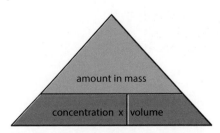

Figure 3.16 Cover the quantity needed to find the formula to use.

DID YOU KNOW?

In Figure 3.15 we have used a few particles to represent a model for concentration. In 100 cm³ of solution there will be huge numbers of particles, possibly at least 100 000 000 000 000 000 000 000!

KEY INFORMATION

1 dm³ is commonly called 1 litre but we use 1 dm³ when using units in scientific contexts.

KEY CONCEPT

Amounts in chemistry

Learning objectives:

- use atomic masses to calculate formula mass
- explain how formula mass relates to number of moles
- explain how number of moles relate to other quantities.

KEY WORDS
....
mole
molar mass
Avogadro's constant

We have learned that there are different ways of measuring chemical quantities. The amount of chemical substance is known as the mole. We have learned how each quantity relates to the mole. Let's draw these altogether here in one place.

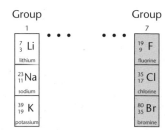

The periodic table shows the atomic mass of each element.

Formula mass

The atomic mass of an element is written on the periodic table.

A formula mass is the sum of the atomic masses of elements in a substance.

The subscript $_3$ means three atoms in the molecule, for example:

$$CaCO_3$$

The formula mass (M_r) is equal to the sum of the atomic masses (A_r) of the five atoms of the three elements: one Ca atom, one C atom and three O atoms.

$$M_r \text{ of } CaCO_3 = 40 + 12 + (3 \times 16)$$
$$= 40 + 12 + 48$$
$$= 100$$

If there is more than one atom in a group attached to another atom a bracket is used, for example:

$$Ca(OH)_2$$

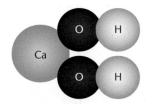

1 Calculate the formula mass of magnesium sulfate, $MgSO_4$.

2 Calculate the formula mass of (a) $Ca(NO_3)_2$ and (b) $Al_2(SO_4)_3$.

HIGHER TIER ONLY

Numbers of moles

The amount of substance of a chemical is the **mole**.

This relates to other quantities such as the **molar mass** and concentration.

In summary, the mole is an amount of substance that:

- contains the molar mass in grams
- has the **Avogadro's** constant of particles (6.02×10^{23})
- relates to the molar mass in grams dissolved in 1 dm^3 of solution (1000 cm^3)
- reacts with other substances in whole number ratios that can be written as an equation, for example,

DID YOU KNOW?

We have seen the detail of these other quantity measurements in other pages in this chapter.

2 moles NaOH : 1 mole H_2SO_4

2 NaOH + H_2SO_4 → Na_2SO_4 + $2H_2O$

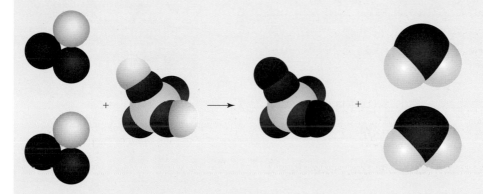

3. **Work out the molar mass of NaOH.**

4. **Calculate the number of particles that two moles of neon contains.**

Working with moles

The key to working with moles is the chemical equation. The balanced symbol equation indicates the number of moles that are involved in any reaction. For example,

$$ZnCO_3 → ZnO + CO_2$$

1 mole of $ZnCO_3$ produces 1 mole of ZnO + 1 mole of CO_2

This is related to the molar masses:

M_r of $ZnCO_3$ = 125 M_r of ZnO = 81 M_r of CO_2 = 44

So 125 g of $ZnCO_3$ will produce 81 g of ZnO and 44 g of CO_2

5. **How many grams of ZnO would be produced from 6.25 g $ZnCO_3$?**

6. **Calculate the mass of CO_2 produced when 660 g of propane, C_3H_8 is combusted.**

$$C_3H_8 + 5O_2 → 3CO_2 + 4H_2O$$

MATHS SKILLS

Change the subject of an equation

Learning objectives:

- to use an equation to demonstrate conservation
- to change the subject of an equation
- to carry out a multi-step calculation.

KEY WORDS

equation
multi-step calculation
multiply

If we are doing a titration or a rate of reaction investigation we will need to measure quantities. Sometimes this requires a *multi-step calculation*. In this process you will often need to change the subject of an equation. There are several ways to do this. Do you have a way of remembering how to change the equations?

Conservation of mass

If we use a total mass of reactants in a chemical reaction the products made will be of the same total mass. This is known as the conservation of mass. In this reaction:

magnesium oxide + sulfuric acid → magnesium sulfate + water

$$MgO \quad + \quad H_2SO_4 \quad → \quad MgSO_4 \quad + \quad H_2O$$

40 g of MgO will make 120 g of $MgSO_4$ + 18 g of H_2O

Why do we get the bigger mass of product? It is because the mass of sulfuric acid has to be added in.

This is because:

mass of MgO + mass of H_2SO_4 = mass of $MgSO_4$ + mass of H_2O

What mass of sulfuric acid will be used?

40 + mass of H_2SO_4 = 120 + 18

The subject of the **equation** needs to be the mass of sulfuric acid. In order to isolate the mass of the sulfuric acid, we need to subtract the mass of the MgO.

As the equation is balanced we need to do the *same to both sides* to move the 40 g.

Therefore we need to *subtract 40 g from both sides*.

(40 – 40) + mass of H_2SO_4 = 120 + 18 – 40

mass of H_2SO_4 = 98

KEY INFORMATION

You will need to know how to change the subject of an equation by using all four operations: add, subtract, **multiply** and divide.

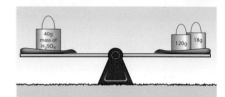

Figure 3.17

1. Use the equation of the reaction to work out how much $MgSO_4$ is made from 24.5 g of H_2SO_4 and 10 g of MgO, when 4.5 g water is made. Write out your working to show how you change the subject of the equation.

2. Use the equation $MgCO_3$ + H_2SO_4 → $MgSO_4$ + H_2O + CO_2 to find out how much $MgCO_3$ was needed to react with 49 g of H_2SO_4 to make 91 g of the total product. Write out your working to show how you change the subject of the equation.

Calculating the mass needed

If we are trying to make a product like Epsom bath salts we need to know how much reactant to start with. These salts are magnesium sulfate, which we can make in two ways:

$$MgO + H_2SO_4 \rightarrow MgSO_4 + H_2O$$

or

$$MgCO_3 + H_2SO_4 \rightarrow MgSO_4 + H_2O + CO_2$$

We can use the relative atomic mass (A_r) to calculate the relative formula mass (M_r) of any of the substances in this equation and so find a ratio of how much reactant gives a quantity of product.

For example: A_r: Mg is 24, O is 16, S is 32, H is 1
so: M_r: MgO is 40 M_r: MgSO$_4$ is 120

We can therefore say that 40 g of MgO will give 120 g of MgSO$_4$. So the ratio is expressed as:

$$\frac{\text{Mass of MgO}}{\text{Mass of MgSO}_4} = \frac{40}{120}$$

If we use 2.0 g MgO, to find out how much MgSO$_4$ we can make, we will need to change the subject of the ratio equation by keeping it balanced at all times:

$$\frac{2}{\text{mass of MgSO}_4} = \frac{40}{120} \qquad \text{Step 1 \textbf{multiply} through by the mass of MgSO}_4$$

$$\frac{2 \times \text{mass of MgSO}_4}{\text{mass of MgSO}_4} = \frac{40 \times \text{mass of MgSO}_4}{120} \qquad \text{Step 2 cancel the fractions}$$

$$2 = \frac{40 \times \text{mass of MgSO}_4}{120} \qquad \text{Step 3 multiply both sides by 120}$$

$$2 \times 120 = \frac{40 \times \text{mass of MgSO}_4 \times 120}{120} \qquad \text{Step 4 cancel the fraction}$$

$$2 \times 120 = 40 \times \text{mass of MgSO}_4 \qquad \text{Step 5 divide both sides by 40}$$

$$\frac{2 \times 120}{40} = \frac{40 \times \text{mass of MgSO}_4}{40} \qquad \text{Step 6 cancel the fractions to isolate the mass of the MgSO}_4$$

$$\frac{2 \times 120}{40} = \text{mass of MgSO}_4 \qquad \text{Step 7 calculate the left hand side}$$

$$6 = \text{mass of MgSO}_4 \qquad \text{Step 8 Is this a reasonable answer? Yes}$$

3 What is the mass of MgSO$_4$ produced from 62.5 g of MgO?

4 What is the mass of MgSO$_4$ produced from 25.2 g of MgCO$_3$?

DID YOU KNOW?

You can do steps 1–5 all at the same time.

Check your progress

You should be able to:

state the law of the conservation of mass → explain how to balance equations in terms of numbers of atoms on both sides of the equation → explain the meaning of subscripts within a formula and multipliers before a formula in a balanced equation

be able to calculate a relative formula mass from the sum of the relative atomic masses → calculate the sum of the relative formula masses of reactants and products → show how the relative formula masses of reactants are equal to the relative formula masses of products

explain that when there is a mass change in a reaction it may be because a gas is being given off → explain why there appears to be a mass change when metal carbonates are heated or metals are heated in oxygen → explain observed changes in mass in non-enclosed systems and explain the changes in terms of the particle model

describe that whenever a measurement is made there is always a degree of uncertainty about the result → represent a distribution of results and make estimates of uncertainty → represent the range of a set of measurements about a mean as a measure of uncertainty

describe the measurement of amounts of substance in moles → calculate the number of moles in a given mass → calculate the mass of a given number of moles

calculate the masses of substances in a balanced symbol equation → calculate the masses of reactants and products from balanced symbol equations → calculate the mass of a given reactant or product

convert masses in grams to amounts in moles → balance an equation given the masses of reactants and products → change the subject of a mathematical equation

recognise that when a reaction has stopped one of the reactants has been used up → describe the reactant that is used up first in a reaction as the limiting reactant → explain the effect of a limiting quantity of a reactant on the amount of products it is possible to obtain, using moles or grams

relate mass, volume and concentration → calculate the mass of solute in a solution → relate concentration in mol/dm^3 to mass and volume

Worked example

1 **Calculate the formula mass of CO_2. The atomic mass of carbon is 12 and atomic mass of oxygen is 16**

CO_2 is $12 + 16 + 16 = 44$

> The answer is correct, but the sentence should not contain is and =

2 **Sam and Alex heat some copper carbonate. They heat 6.17 g.**

After heating and cooling the mass of the black powder left was 4.02 g

a **What was the reason for the decrease in mass?**

A gas was given off

> This answer is correct but could also identify the gas as carbon dioxide

b **What was the black powder formed?**

(copper oxide)   carbon   copper

> copper oxide is correct

c **Write a word equation for the decomposition**

copper carbonate + heat → copper oxide + carbon dioxide

> The chemicals are correct but it is not correct to write heat in the equation

d **Calculate the decrease in mass**

Mass loss $= 6.17 - 4.02$
$= 2.15$

> The mass is correct but there needs to be units of g.

3 **Using the masses in question 2, calculate**

a **How many moles of copper carbonate, $CuCO_3$, Sam and Alex heated?**

A_r Cu = 63.5 A_r C = 12 A_r O =16

Relative formula mass of $CuCO_3$ $= 63.5 + 12 + (3 \times 16)$
$= 123.5$

Number of moles $= \dfrac{6.17}{123.5} = 0.05$ moles

> The relative formula mass and calculation are correct.

b **How many moles of black powder was produced?**

Relative formula mass of CuO $= 63.5 + 16 = 79.5$

Number of moles $= \dfrac{4.02}{79.5} = 0.05$ moles

> The relative formula mass and calculation are correct.

c **Calculate the mass of CO_2 that was given off using the number of moles that were made.**

Number of moles of $CO_2 = 0.05$ moles
Relative formula mass of $CO_2 = 44$
Mass of CO_2 formed $= 0.05 \times 44 = 2.2$

> The relative formula mass and calculation are correct, but the units of mass need to be included.

End of chapter questions

Getting started

1 Calculate the formula mass of $NaNO_3$. Atomic masses: Na=23 N=14 O=16

Circle your answer.   53   113   85   `1 Mark`

2 Alex burns a 0.7 g strip of magnesium in air. What will be the most likely mass after heating?

a 0.7 g   **b** 0.64 g   **c** 0.76 g   `1 Mark`

3 Sam heats 4.2 g of a substance. After heating the mass is 3.8 g. Suggest why.   `2 Marks`

4 Hydrochloric acid is added to magnesium pieces in a flask. After 70 seconds no more hydrogen was given off and there were no more magnesium pieces. Which chemical is the limiting factor?

a hydrochloric acid   **b** magnesium   **c** hydrogen   `1 Mark`

5 Calculate the relative formula mass of $Ca(OH)_2$. A_r Ca = 40 A_r O = 16 A_r H = 1   `1 Mark`

6 Explain why mass is conserved in this reaction. (A_r Na =23 A_r O = 16).

$$4Na + O_2 \rightarrow 2Na_2O$$ `2 Marks`

7

Atom	Number of protons	Number of neutrons
X	22	23
Y	22	24
Z	23	24

Calculate and explain which atom has the highest relative atomic mass.   `2 Marks`

Going further

8 Calculate the relative atomic mass of element X which has two isotopes, atomic mass 7 and 8, where the relative abundance of the two isotopes is 50% of each.   `1 Mark`

9 Both of a pair of isotopes have the same:

a number of protons   **b** number of neutrons   **c** atomic mass   `1 Mark`

10 Explain which standard is used when measuring relative atomic masses.   `2 Marks`

11 Use the data table in question 7.

a Explain which two atoms are isotopes.

b Write the formula of Z with the proton number and the mass number identified.

c Write the formula of an isotope of Z that has a mass number of 49.   `4 Marks`

12 Calculate the number of moles in 10 g of MgO.   `2 Marks`

More challenging

13 What is the number of particles in one mole of a substance?

    a 6.02×10^{23} **b** 6.02×10^{4} **c** 6.02×10^{-9} **1 Mark**

14 When the concentration of a solution = mass of solute/ volume of solvent, the units are: **a** $moles/cm^3$ **b** $g/mole$ **c** g/dm^3 **1 Mark**

15 Explain the meaning of 'percentage uncertainty'.   **2 Marks**

16 Explain the term 'limiting factor'.   **2 Marks**

17 Using the atomic masses on the periodic table, complete the table below and work out the molecular formula of the hydrocarbon C_nH_{2n+2}

Formula	Mass in grams	Molar mass	Number of moles
$MgCO_3$	42		
$CaCO_3$			0.2
$Ca(OH)_2$	148		
C_nH_{2n+2}	25		0.25

4 Marks

Most demanding

18 Explain the meaning of Avogadro's constant.   **2 Marks**

19 $Mg + 2HCl \rightarrow MgCl_2 + H_2$

    a Identify how many moles of hydrochloric acid react with one mole of magnesium.

    b Calculate the mass of 4 moles of Mg.

    c Calculate the mass of magnesium chloride that is made from 1.2 g of Mg.

    A_r Mg = 24 A_r Cl = 35.5 A_r H =1   **4 Marks**

20 Hydrocarbons burn in oxygen (O_2) to make CO_2 and H_2O.

Hydrocarbon	Number of moles of oxygen needed	Mass of CO_2 made/g
C_5H_{12}	9	220
C_9H_{20}	13	396

    A_r C = 12 A_r O = 16 A_r H = 1

    a Calculate the formula mass of CO_2

    b Calculate the number of moles of CO_2 made by each hydrocarbon.   **4 Marks**

Total: 40 Marks

CHEMICAL CHANGES

IDEAS YOU HAVE MET BEFORE:

REACTIVITY SERIES

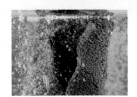

- Magnesium reacts more readily than zinc with acid.
- Copper does not react with dilute acid.
- If a piece of zinc is put into copper sulfate, copper coats the zinc.

EXTRACTION OF METALS

- Gold is dug up as unreacted metal from the ground.
- Heating the green ore, malachite, turns it to copper oxide.
- Furnaces for melting iron were very important for making steel.

NEUTRALISATION AND SALT PRODUCTION

- Acids turn universal indicator red.
- Alkalis turn universal indicator blue.
- Copper sulfate solution can be left to make blue crystals.

STRONG AND WEAK ACIDS

- Lemons and vinegar are acids found at home.
- Hydrochloric acid reacts with marble chips to make CO_2.
- Sulfuric acid can react with magnesium to make hydrogen.

ELECTROLYSIS

- Copper and graphite conduct electricity.
- Copper put into silver nitrate makes a coat of silver.
- Transformers can change the voltage from 240 V a.c. to 12 V d.c.

IN THIS CHAPTER YOU WILL FIND OUT ABOUT:

WHY ARE SOME METALS MORE REACTIVE THAN OTHERS?

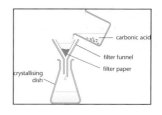

- Metals from Group 1 show reactivity with water.
- Other metals react with acids with varying degrees of activity.
- More reactive metals displace less reactive metals.

WHY ARE SOME METALS EXTRACTED BY REDUCTION WITH CARBON?

- Less reactive metals can be extracted by reduction with carbon.
- Reduction is the removal of oxygen, oxidation is gaining oxygen.
- Oxidation and reduction can be represented by ionic equations.

HOW DO ACID AND BASES PRODUCE NEUTRAL SALTS?

- Acids react with bases or alkalis to produce salts and water.
- Acids react with carbonates to make salts, water and carbon dioxide.
- Solid reactants are added in excess to acids then filtered off.

WHY ARE SOME ACIDS STRONG AND OTHER ACIDS WEAK?

- pH is the concentration of hydrogen ions and measures acidity.
- Strong acids fully ionise in solution and have low pH.
- Weak acids partially ionise and have slightly higher pH.

WHY ARE SOME METALS PRODUCED BY ELECTROLYSIS?

- Molten compounds allow ions to move, so conduct electricity.
- Aluminium oxide melts and electrolyses to produce aluminium.
- Electrolysis of solutions often produces hydrogen and oxygen.

Metal oxides

Learning objectives:

- identify that metals react with oxygen to form metal oxides
- explain oxidation by gain of oxygen
- identify metal oxides as bases.

KEY WORDS

metal oxide
ore
oxidation
reduction

We use metals such as iron, copper, tin and lead on a large scale for tools, implements and machinery. However, most metals are found combined as minerals in ores. Many of these minerals are oxides, and over the centuries, methods have been developed to extract them. More recently, aluminium has also been extracted. All these metals react with oxygen to produce *metal oxides*.

Iron(III) oxide

Iron is the most abundant element on Earth by mass and the fourth most abundant element in the Earth's crust. Iron can be processed into steel. Steel is one of the most important materials used on a global scale, second only to crude oil.

Iron does not exist naturally on its own, but combined with other compounds as minerals. These minerals form **ores** that can be dug out of the ground. Iron ore is mined as a red-brown rock. Some of the largest mines are in Australia and Brazil.

One common ore of iron is called haematite. (Notice the same word root in *haem*oglobin). This ore contains iron oxide.

Iron oxide is made when iron reacts with oxygen.

Figure 4.1 Haematite being mined in Australia. Iron(III) oxide ores are red-brown.

$$\text{iron} + \text{oxygen} \rightarrow \text{iron oxide}$$

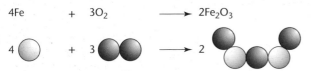

$$4\text{Fe} + 3\text{O}_2 \rightarrow 2\text{Fe}_2\text{O}_3$$

Figure 4.2

This is an **oxidation** reaction because the metal *gains* oxygen.

Iron(III) oxide can be **reduced** by carbon monoxide. It is reduced because it *loses* oxygen.

$$\text{iron(III) oxide} + \text{carbon monoxide} \rightarrow \text{iron} + \text{carbon dioxide}$$

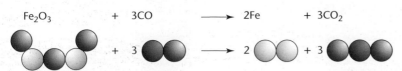

$$\text{Fe}_2\text{O}_3 + 3\text{CO} \rightarrow 2\text{Fe} + 3\text{CO}_2$$

Figure 4.3

KEY INFORMATION

Iron can form two types of compounds, iron(II) compounds, which are often a grey/green colour and iron(III) compounds, which are often a red/brown colour.

1. Explain what type of reaction occurs if zinc reacts with oxygen.
2. Explain what type of reaction occurs if lead oxide reacts with carbon.

Aluminium oxide

Aluminium is the third most abundant element in the Earth's crust (behind oxygen and silicon) and the most abundant metal.

An important aluminium ore is bauxite, aluminium oxide.

Aluminium metal is very useful as it is less dense than many other metals. One of its many uses is in aircraft.

A thin film of aluminium oxide forms on the surface of aluminium metal and protects it. The aluminium does not corrode further with oxygen.

$$aluminium + oxygen \rightarrow aluminium\ oxide$$

Aluminium oxide cannot be reduced by carbon, as the metal is too reactive. The aluminium oxide ore has to be reduced by electrolysis (see topic 4.14).

3. Explain why aluminium is more expensive to extract from its ore than iron is.

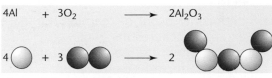

$$4Al + 3O_2 \longrightarrow 2Al_2O_3$$

Figure 4.4

Figure 4.5 White sodium oxide powder

Oxides of Group 1 and Group 2

Sodium oxide is formed when sodium is burned in oxygen. It is a white powder and is soluble in water.

During the oxidation reaction, the sodium atoms lose electrons to make positive ions. The electrons are transferred to the oxygen atoms, which become negative ions. (See Figure 4.6.)

$$4Na - 4e^- \rightarrow 4Na^+ \qquad 2O + 4e^- \rightarrow 2O^{2-}$$

The dot and cross diagram for this combination is shown in Figure 4.7.

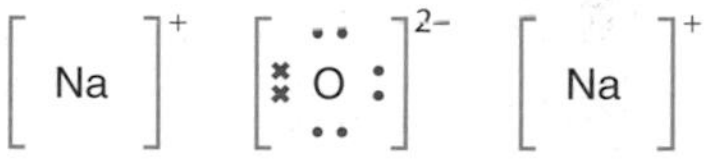

The oxide formed is a basic oxide. When dissolved in water the oxide forms a hydroxide solution. A soluble base is an alkali that makes an alkaline solution.

$$Na_2O + H_2O \rightarrow 2NaOH$$

This solution turns universal indicator purple and has a pH of 14.

Calcium oxide behaves in the same way and also dissolves to produce an alkaline solution.

4. Draw the dot and cross diagram for calcium oxide, CaO.

5. Hydroxide ions have the formula OH⁻. Write the balanced equation for the formation of calcium hydroxide when solid calcium oxide is added to water. Include state symbols in your answer.

$$4Na - 4e^- \longrightarrow 4Na^+ \qquad 2O + 4e^- \longrightarrow 2O^{2-}$$

Figure 4.6

Figure 4.7 Dot and cross diagram for Na_2O

KEY INFORMATION

Four electrons need to be transferred as there are two atoms of oxygen required for the reaction and each atom requires two electrons to become a negative ion.

HIGHER TIER ONLY

6. Write the equation to show the transfer of electrons when calcium is oxidised.

Reactivity series

Learning objectives:

- describe the reactions, if any, of metals with water or dilute acids
- deduce an order of reactivity of metals based on experimental results
- explain how the reactivity is related to the tendency of the metal to form its positive ion.

KEY WORDS

displacement
positive ion
reactivity
tendency

You have already seen that Group 1 metals, sodium and potassium, react vigorously with water. Potassium is more reactive than sodium. In your class, you will have seen a tiny amount of potassium skimming on the surface of water and bursting into a lilac flame. Caesium is even more reactive than potassium.

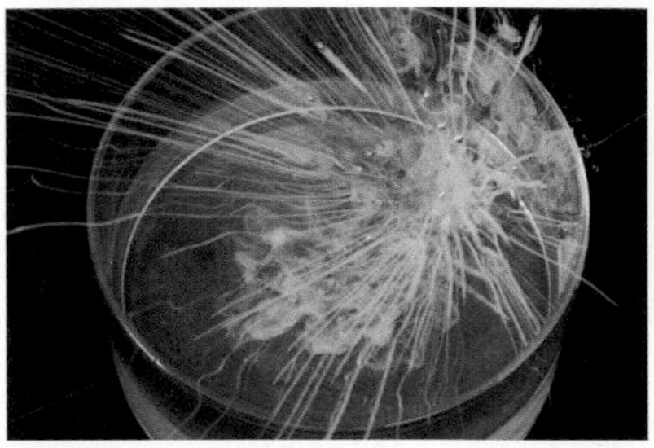

Reactions of metals with water

- Potassium, sodium and lithium react vigorously with water.
- Calcium reacts with water less vigorously than lithium.
- Magnesium reacts with steam but not with cold water.

We can put these five metals into the order of their **reactivity** with water.

    potassium > sodium > lithium > calcium > magnesium

Other, less reactive, metals do not react with water but do react with acid, so the order of reactivity can be continued.

Magnesium, zinc, iron and copper can be added to solutions of acid and their reactivity observed. Copper does not react with dilute acids but magnesium does.

The order of reactivity of metals with acids is:

    magnesium > zinc > iron > copper

So from the two sets of observations the **reactivity** series so far is: potassium > sodium > lithium > calcium > magnesium > zinc > iron > copper

1 **Explain how you know that sodium is more reactive than zinc.**

Positive ions

The metals potassium, sodium, lithium, calcium, magnesium, zinc, iron and copper all form **positive ions**.

They can be put in order of their reactivity by observation of their reactions with water and dilute acids. The reason that there is a difference in reactivity is because of the different **tendency** of each metal to form positive ions.

Figure 4.8 Potassium on water. A highly reactive metal. Don't try this at home.

DID YOU KNOW?

Gold is so unreactive that it does not react with oxygen. It is always beautifully shiny.

Figure 4.9 Gold can remain untarnished and shiny for millennia, as it is so unreactive.

Figure 4.10 Sodium atom forming a sodium ion by losing an electron

Sodium forms **positive ions** much more easily than copper and more easily than lithium.

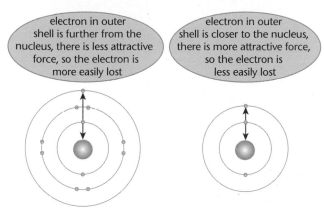

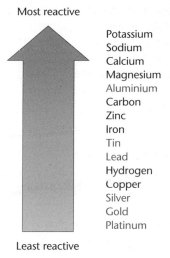

Most reactive

Potassium
Sodium
Calcium
Magnesium
Aluminium
Carbon
Zinc
Iron
Tin
Lead
Hydrogen
Copper
Silver
Gold
Platinum

Least reactive

Figure 4.12 The reactivity series

Figure 4.11 Sodium forms ions more easily than lithium.

The non-metals hydrogen and carbon are also often included in the reactivity series. Hydrogen is placed between iron and copper. Carbon lies between magnesium and zinc.

A more reactive metal (as the metal) can displace a less reactive metal from a solution of a compound of the less reactive metal.

For example, if magnesium metal is put into iron sulfate solution, the magnesium 'pushes the iron out' of the iron sulfate. Solid iron and a solution of magnesium sulfate are made.

This is called a **displacement** reaction.

iron sulfate + magnesium → magnesium sulfate + iron

It happens because magnesium is more reactive than iron. The iron has been pushed out or displaced. The iron forms a coating on the piece of magnesium.

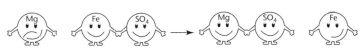

Figure 4.14

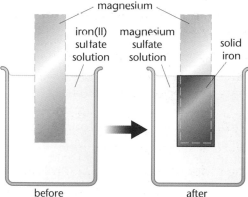

Figure 4.13 Why can magnesium displace iron?

2 **Explain why potassium forms positive ions more easily than sodium.**

HIGHER TIER ONLY

Ion formation

Positive ions are formed by the loss of electrons.

$$K - e^- \rightarrow K^+ \; [\text{ or } K \rightarrow K^+ + e^-]$$

$$Mg - 2e^- \rightarrow Mg^{2+} \; [\text{ or } Mg \rightarrow Mg^{2+} + 2e^-]$$

Displacement of a less reactive metal is represented by:

$$Mg + Fe^{2+} \rightarrow Mg^{2+} + Fe$$

3 **Write an ionic equation to show the displacement of copper ions by zinc metal. Both copper and zinc ions have a 2+ charge.**

KEY SKILLS

Before you start to write equations find out how many electrons the metals lose to make an ion.

Extraction of metals

Learning objectives:

- identify substances reduced by loss of oxygen
- explain how extraction methods depend on metal reactivity
- interpret or evaluate information on specific metal extraction processes.

KEY WORDS
....
reactivity
reduction
reduction with
  carbon

There has been a rapid rise in the need for iron to make steel due to urbanisation in the last decade. Which country is driving this need for more metal extraction?

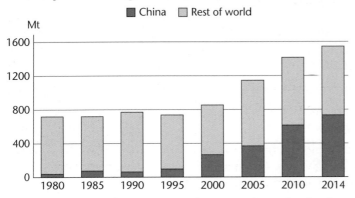

Figure 4.15 Graph of increase in steel consumption. What has happened to consumption by China?

Figure 4.16 Haematite, Fe_2O_3, is the common ore of iron.

Reduction

Metals such as gold are found in the Earth as the metal itself because it is unreactive. Most metals, however, are found as compounds because they react with other elements. Chemical reactions are needed to extract the metal. The reactions needed depend on the **reactivity** of the metal.

Metals more reactive than carbon need to be extracted by electrolysis.	For example:	potassium, sodium, calcium, magnesium, aluminium
Metals less reactive than carbon can be extracted from their oxides by **reduction with carbon**.	For example:	zinc iron lead

Figure 4.17 Zinc blende, ZnS, is the common ore of zinc.

Reduction involves the loss of oxygen.

In the reactions needed to extract zinc, two stages are needed.

a to convert the ore zinc blende (ZnS) to zinc oxide
b to convert the zinc oxide to zinc.

The second stage is the **reduction** of zinc oxide using carbon.

zinc oxide + carbon → zinc + carbon dioxide

1. In the second stage of extracting, zinc oxide is changed to zinc as in the equation above. State which substance is oxidised and which is reduced.

Reduction using carbon

To extract iron…

Iron ore is reduced, which means taking away oxygen.
This is done with a reducing agent.
The reducing agent used is carbon monoxide.
Carbon monoxide is made by heating carbon.

In industry the process has many steps but put simply:

carbon + oxygen → carbon monoxide

iron oxide + carbon monoxide → iron + carbon dioxide

$$2C + O_2 \rightarrow 2CO \qquad\qquad (i)$$

$$Fe_2O_3 + 3CO \rightarrow 2Fe + 3CO_2 \qquad (ii)$$

2. In equation (i) explain which substance has been oxidised.

3. In equation (ii) explain which substance has been oxidised and which substance has been reduced.

4. The ore cuprite contains copper(II) oxide (CuO) and can be smelted with carbon. The ore is mixed with coal and roasted in a furnace. Copper metal and carbon monoxide, CO, are produced.

   a. Write a balanced equation for the reaction.

   b. State what is happening to the copper(II) oxide and carbon.

HIGHER TIER ONLY

Ionic equations

In the extraction of zinc, the equations are:

zinc sulfide + oxygen → zinc oxide + sulfur dioxide

then

$$2ZnO + C \rightarrow 2Zn + CO_2$$

The half equation for zinc ions converting to zinc is:

$$2Zn^{2+} + 4e^- \rightarrow 2Zn$$

In the extraction of iron, the half equation for the final reaction (ii) is:

$$2Fe^{3+} + 6e^- \rightarrow 2Fe$$

5. Write a full balanced symbol equation for the conversion of zinc sulfide, ZnS, to zinc oxide, ZnO, and sulfur dioxide, SO_2.

6. The iron half equation is derived from reaction (ii) for the extraction of iron. Explain this half equation.

REMEMBER!

You do not need to know the details of zinc extraction or details of the processes used in the extraction of iron. You need to know about oxidation and reduction.

Figure 4.18 Molten iron extracted from iron ore

KEY INFORMATION

This is done by reacting the iron ore 'haematite', which is iron oxide, with carbon monoxide.

DID YOU KNOW?

Most iron extracted is turned into steel, by adding carbon and other elements. Steel, an alloy, is less easily corroded.

Oxidation and reduction in terms of electrons

Learning objectives:

- use experimental results of displacement reactions to confirm the reactivity series
- write ionic equations for displacement reactions
- identify in a half equation which species are oxidised and which are reduced.

KEY WORDS

displacement
half equations
ionic equations
oxidation
reduction

As we have seen, a more reactive metal can displace a less reactive metal from a solution of its compound. This is called a displacement reaction. These reactions are used to save ships from rusting and rely on one metal's ability to transfer its electrons to another metal.

Experimental results

An example of a **displacement** reaction is:

iron sulfate + magnesium → magnesium sulfate + iron

It happens because magnesium is more reactive than iron. The iron has been pushed out or displaced. It forms a coating on the rest of the magnesium (see Figure 4.13 in topic 4.2).

This table shows what happens when four different metals are used.

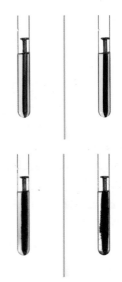

Figure 4.19 A displacement reaction. Iron is reacting with copper sulfate. What can you see is happening?

Solution used	Metal being added			
	magnesium	zinc	iron	copper
magnesium sulfate	X	X	X	X
zinc sulfate	✓	X	X	X
iron sulfate	✓	✓	X	X
copper sulfate	✓	✓	✓	X

Key:

X   means that nothing happens

✓   means that the added metal gets coated with the metal from the solution

The table shows that copper does not react with any of the other solutions and is, therefore, the least reactive of these metals.

1. **Use the experimental results in the table to put the metals in order of reactivity.**

2. **When a piece of tin metal is placed in three different solutions this is seen:**

copper sulfate: tin is coated	zinc sulfate: nothing happens	iron(II) sulfate: nothing happens

Explain the position of tin in the reactivity series.

KEY SKILL

To work out what happens in a displacement reaction you do not need to remember every reaction. You need to know the order of reactivity: magnesium, zinc, iron, copper.

This means that:

- magnesium metal displaces zinc, iron and copper
- zinc displaces iron and copper
- iron displaces copper.

Displacement equations

When writing an equation for a displacement reaction, the more reactive metal swaps places with the less reactive metal. (It does not matter whether the compound is a chloride, a nitrate or a sulfate.)

For example, as zinc is more reactive than iron, iron is displaced from the solution.

$$Zn + FeCl_2 \rightarrow ZnCl_2 + Fe$$

3 Write an equation for the displacement reaction between nickel sulfate ($NiSO_4$) and the more reactive metal, magnesium.

DID YOU KNOW?

The half equation $Zn - 2e^- \rightarrow Zn^{2+}$ is conventionally written as

$$Zn \rightarrow Zn^{2+} + 2e^-$$

but with the first way it is easier to see what is happening.

HIGHER TIER ONLY

Ionic and half equations

All metals react by losing electrons and turning into ions. This is **oxidation**.

Displacement reactions depend on having a reactive metal element and the compound of a less reactive metal.

The more reactive the metal, the harder it pushes off electrons.

These electrons have to go somewhere; they are forced onto the ions of other metals that are not so reactive. The metal atoms that gain these electrons are reduced. This is **reduction**.

These **ionic equations** show what happens:

$$Mg + Zn^{2+} \rightarrow Mg^{2+} + Zn$$
$$Zn + Fe^{2+} \rightarrow Zn^{2+} + Fe$$

These equations can also be written as **half equations** to show the electron transfer happening.

$$Mg - 2e^- \rightarrow Mg^{2+} \qquad Zn^{2+} + 2e^- \rightarrow Zn$$
$$Zn - 2e^- \rightarrow Zn^{2+} \qquad Fe^{2+} + 2e^- \rightarrow Fe$$

oxidation is the loss of electrons and reduction is the gain of electrons.

KEY INFORMATION

Oxidation Is electron Loss
Reduction Is electron Gain:
OILRIG

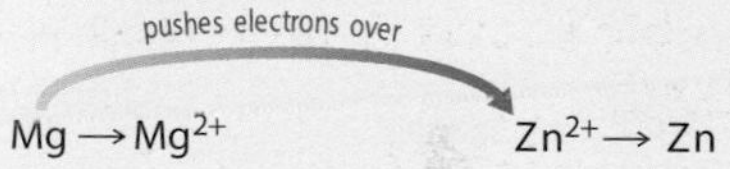

pushes electrons over

$Mg \rightarrow Mg^{2+}$ $Zn^{2+} \rightarrow Zn$

Figure 4.20

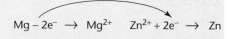

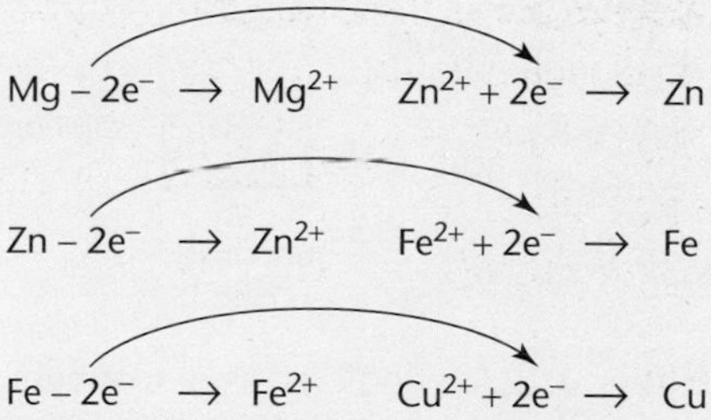

$Mg - 2e^- \rightarrow Mg^{2+}$ $Zn^{2+} + 2e^- \rightarrow Zn$

$Zn - 2e^- \rightarrow Zn^{2+}$ $Fe^{2+} + 2e^- \rightarrow Fe$

$Fe - 2e^- \rightarrow Fe^{2+}$ $Cu^{2+} + 2e^- \rightarrow Cu$

Figure 4.21

4 Write an ionic equation for the reaction between magnesium metal and copper chloride solution.

5 Write two half equations for the reaction between magnesium metal and iron(II) chloride solution.

6 Identify in these two half equations which species is oxidised and which is reduced.
$Al - 3e^- \rightarrow Al^{3+}$ $Cr^{3+} + 3e^- \rightarrow Cr$

7 Magnesium is more reactive than silver. It can displace silver ions, Ag^+, from solution.

  a Write the half equations for the displacement of silver ions by magnesium. Identify which species is oxidised and which is reduced.

  b Write the ionic equation for the displacement reaction. The number of electrons transferred must be equal.

Reaction of metals with acids

Learning objectives:

- describe how to make salts from metals and acids
- write full balanced symbol equations for making salts
- use half equations to describe oxidation and reduction.

Crystals of salts can be made in many ways, including reacting metals with acids. To make a different salt, you use a different acid.

Making salts

Acids react with some **metals** to produce **salts** and hydrogen.

Magnesium metal reacts fairly vigorously with dilute sulfuric acid to make a solution of magnesium sulfate and hydrogen.

magnesium + sulfuric acid → magnesium sulfate + hydrogen

To make a crystallised salt:

a excess magnesium needs to be added to the acid
b the solution needs to be filtered into a crystallising dish
c the solution needs to be concentrated by evaporation
d and the solution then left to evaporate, to crystallise.

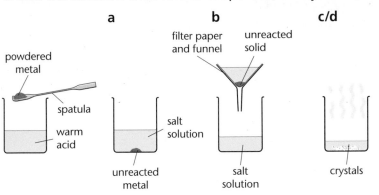

Figure 4.22 Making crystals of magnesium sulfate

Salts can also be made with zinc and iron. For example,

zinc + hydrochloric acid → zinc chloride

1 Write the name of the salt made from magnesium and hydrochloric acid.

2 Write the word equation for the formation of iron(II) sulfate.

Forming ions

Chemical equations for the reaction of metals with acids can be constructed using symbols of metals and formulae of acids:

> **KEY INFORMATION**
>
> Don't forget that a salt is the neutral product of an acid added to a base or a carbonate. It is made of two ions: the positive ion from the base or carbonate and the negative ion from the acid. Sodium chloride is only one example; there are many, many others.

> **KEY INFORMATION**
>
> When making salts, hydrochloric acid makes chlorides and sulfuric acid make sulfates. Iron can make iron(II) salts or iron(III) salts. They are usually different colours.

Metals		Acids			
magnesium	Mg^{2+}	sulfuric acid	H_2SO_4	$2H^+$	SO_4^{2-}
iron	Fe^{2+}	hydrochloric acid	HCl	H^+	Cl^-
zinc	Zn^{2+}				

In making magnesium sulfate a 2+ ion is joining with a 2– ion, so the formula is a 1:1 relationship, $MgSO_4$.

magnesium + sulfuric acid → magnesium sulfate + hydrogen

$$Mg + H_2SO_4 \rightarrow MgSO_4 + H_2$$

In making magnesium chloride a 2+ ion is joining with a 1– ion. This means that two of the 1– ions are needed, so the formula is a 1:2 relationship, $MgCl_2$.

magnesium + hydrochloric acid → magnesium chloride + hydrogen

$$Mg + 2HCl \rightarrow MgCl_2 + H_2$$

As two chloride ions (Cl^-) are needed to join with one Mg^{2+} ion, they must both be provided by the HCl. Therefore, two amounts of HCl are needed. The equation is balanced.

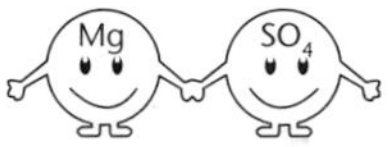

3 Write a full balanced symbol equation for the production of zinc chloride from zinc.

4 Write a full balanced symbol equation for the production of iron(III) chloride, $FeCl_3$, from iron.

Figure 4.23

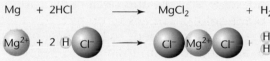

Figure 4.24

Mg + 2HCl ⟶ $MgCl_2$ + H_2

Mg^{2+} + 2 H Cl^- ⟶ Cl^- Mg^{2+} Cl^- + H H

Figure 4.25

Using half equations

When making a magnesium salt from magnesium, this reaction happens:

$$Mg + 2H^+ \rightarrow Mg^{2+} + H_2$$

The **half equations** are:

$Mg - 2e^- \rightarrow Mg^{2+}$	$2H^+ + 2e^- \rightarrow H_2$
Electron loss is oxidation	Electron gain is reduction
As both these reactions take place at the same time this is a *redox* reaction	

In the equation:

- Magnesium loses electrons so oxidation takes place to produce magnesium ions (Mg^{2+}).
- Hydrogen ions (H^+) gain electrons so reduction takes place.

5 Write two half equations for the reaction between zinc and an acid.

6 In the reaction between iron and sulfuric acid, which component is oxidised and which is reduced?

7 Write a balanced chemical equation for the reaction of aluminium with sulfuric acid. Write half equations for this reaction. Aluminium forms Al^{3+} ions.

Neutralisation of acids and salt production

Learning objectives:

- describe ways that salts can be made
- predict products from given reactants
- deduce the formulae of salts from the formulae of common ions.

KEY WORDS

alkali
base
carbonate
salt

You may have made copper sulfate crystals before. Salts are made when acids are neutralised by alkalis, bases and metal carbonates. You can work out the name of the salt by using the name from the base followed by the name from the acid.

Figure 4.26 Making copper sulfate crystals. Copper sulfate is a salt.

Alkalis and bases

Metal oxides and metal hydroxides are **bases**. A few bases are soluble in water, and these are called **alkalis**, for example, sodium hydroxide and calcium hydroxide.

When an acid and a base react, *neutralisation* takes place to make a **salt** and water. The word equation for neutralisation is:

acid + base → salt + water

For example, the reaction of sodium hydroxide, a soluble base, with hydrochloric acid produces sodium chloride.

sodium hydroxide + hydrochloric acid → sodium chloride + water

Salts can also be made by reacting acids with excess of an insoluble base.

For example, zinc oxide is an insoluble base. Zinc oxide and sulfuric acid react to make zinc sulfate and water.

zinc oxide + sulfuric acid → zinc sulfate + water

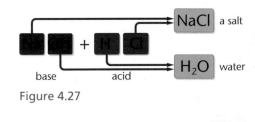

base   acid

Figure 4.27

Figure 4.28 Making a salt with excess base or carbonate

Metal **carbonates** react with dilute acid to make a metal salt, but this time the gas carbon dioxide is produced. Water is also formed.

acid + metal carbonate → salt + water + carbon dioxide

KEY INFORMATION

'Common salt' is sodium chloride, NaCl, but other chlorides, sulfates and nitrates are also salts.

Magnesium carbonate reacts with hydrochloric acid to make magnesium chloride, carbon dioxide and water.

magnesium + hydrochloric → magnesium + carbon dioxide + water
carbonate   acid   chloride

1 **Write the word equation for the reaction between potassium hydroxide and hydrochloric acid.**

2 **Write the word equation for the reaction between copper carbonate and nitric acid.**

Predicting the names of salts

Acids react with alkalis to make a salt and water. The second part of the name of the salt is taken from the acid.

The first part of the name of the salt is taken from the metal in the base.

3 **Predict the name of the salt formed when nitric acid reacts with zinc oxide.**

4 **Predict the name of the salt made from the reaction between sulfuric acid and copper carbonate.**

Common acid	The salt made
nitric	nitrate
sulfuric	sulfate
hydrochloric	chloride

DID YOU KNOW?

The names of salts from other acids can be predicted. For example, citric acid (the acid in lemons) makes sodium citrate.

Formulae of salts from ions

The formula of a salt can be deduced from the formulae of the ions.

This is a table of common ions showing their charges:

Ions from alkalis and carbonates		Ions from bases and carbonates		Ions from acids	
sodium	Na$^+$	magnesium	Mg^{2+}	chloride	Cl$^-$
potassium	K$^+$	zinc	Zn^{2+}	nitrate	NO$_3^-$
calcium	Ca^{2+}	copper	Cu^{2+}	sulfate	SO$_4^{2-}$

For example, potassium hydroxide and nitric acid make potassium nitrate. The formula of the salt can be written by looking at the ions in the table.

K$^+$ single charge,   NO$_3^-$ single charge,   formula KNO$_3$

Magnesium sulfate:

Mg^{2+} double charge,   SO$_4^{2-}$ double charge,   formula MgSO$_4$

Copper chloride:

Cu^{2+} double charge,   Cl$^-$ single charge,   formula CuCl$_2$

Sodium sulfate:

Na$^+$ single charge,   SO$_4^{2-}$ double charge,   formula Na$_2$SO$_4$

5 **Deduce the formula of zinc chloride.**

6 **Deduce the formula of copper nitrate.**

7 **Write a balanced equation for the reaction between calcium carbonate and nitric acid.**

KEY INFORMATION

The nitrate ion will need a bracket around it in this formula. For example, calcium nitrate is Ca(NO$_3$)$_2$

Soluble salts

Learning objectives:

- describe how to make pure, dry samples of soluble salts
- explain how to name a salt
- derive a formula for a salt from its ions.

KEY WORDS

concentrate
crystallise
filter

We have seen how crystals can be made from acids and insoluble substances. Soluble salts, such as bath salts and some fertilisers, can be made very simply in the lab by reacting, filtering, evaporating and crystallising.

Soluble salts

A soluble salt can be made by reacting an acid with an insoluble solid substance.

- The solid is added to the acid until no more of the solid reacts and it is in *excess*.
- The excess solid is **filtered** off leaving a solution of the salt.

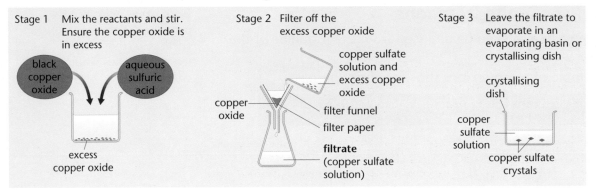

Figure 4.29 Making a salt from an acid and a base or a carbonate

- The salt solution is then heated a little to **concentrate** it.
- The concentrate is left to cool and **crystallise** to produce a solid salt.

1 Explain why an excess of the solid is added to the acid.

2 Explain why the solution that has been filtered needs to be concentrated by heating a little.

DID YOU KNOW?

Crystals are larger if the solution is cooled more slowly.

Which salts can be made?

Acids can make salts by reacting with insoluble substances such as metals, metal oxides, metal hydroxides or metal carbonates.

metal   + acid → metal salt + hydrogen

metal oxide   + acid → metal salt + water

metal hydroxide + acid → metal salt + water

metal carbonate + acid → metal salt + water + carbon dioxide

Some common chemical equations for the neutralisation of acids by a base or metal carbonate can be constructed using these formulae:

Acids		Bases and alkalis		Carbonates		Metals	
sulfuric acid	H_2SO_4	potassium hydroxide	KOH	sodium carbonate	Na_2CO_3	magnesium	Mg
nitric acid	HNO_3	sodium hydroxide	NaOH	calcium carbonate	$CaCO_3$	zinc	Zn
hydrochloric acid	HCl	calcium hydroxide	$Ca(OH)_2$	zinc carbonate	$ZnCO_3$	iron	Fe
		copper oxide	CuO	copper carbonate	$CuCO_3$		

The equation for the reaction:

- between iron and sulfuric acid is $Fe + H_2SO_4 \rightarrow FeSO_4 + H_2$
- between zinc oxide and sulfuric acid is $ZnO + H_2SO_4 \rightarrow ZnSO_4 + H_2O$
- between sodium hydroxide and nitric acid is $NaOH + HNO_3 \rightarrow NaNO_3 + H_2O$
- between copper carbonate and sulfuric acid is $CuCO_3 + H_2SO_4 \rightarrow CuSO_4 + H_2O + CO_2$

3 Write the equation for the reaction between copper oxide and sulfuric acid.

4 Write the equation for the reaction between potassium hydroxide and hydrochloric acid.

Charges on ions

The ions used in the examples above are:

Single charge	Cl^-	NO_3^-	OH^-	Na^+	K^+	H^+
Double charge	SO_4^{2-}	Ca^{2+}	Mg^{2+}	Zn^{2+}	Fe^{2+}	Cu^{2+}

The equations underneath the table are all balanced. This is because all the ions in the reactants and products are matched for charges.

If the salt is made using ions that are +2 and – 1 (for example, copper chloride $CuCl_2$) or using ions that are +1 and – 2 (for example, sodium sulfate Na_2SO_4) then the equations will need to be balanced.

$$CuO + 2HCl \rightarrow CuCl_2 + H_2O$$

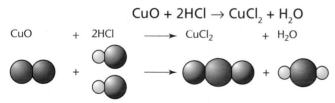

CuO   +   2HCl   ⟶ $CuCl_2$ + H_2O

Figure 4.30

KEY INFORMATION

Remember the nitrate ion will need a bracket around it in this formula, e.g. $X(NO_3)_y$

$$2NaOH + H_2SO_4 \rightarrow Na_2SO_4 + H_2O$$

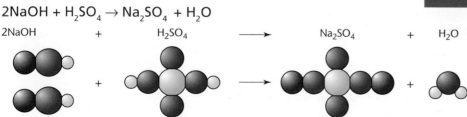

2NaOH   + H_2SO_4 ⟶ Na_2SO_4 + H_2O

Figure 4.31

5 Write a balanced equation for the reaction between zinc carbonate and hydrochloric acid.

6 Write a balanced equation for the reaction between magnesium oxide and nitric acid.

REQUIRED PRACTICAL

Preparing a pure, dry sample of a soluble salt from an insoluble oxide or carbonate

KEY WORDS

carbonate
filter
crystallisation

Learning objectives:

- describe a practical procedure for producing a salt from a solid and an acid
- explain the apparatus, materials and techniques used for making the salt
- describe how to safely manipulate apparatus and accurately measure melting points.

Being able to choose the correct techniques and carry out a specified procedure to produce a pure product is an important skill for scientists. You will probably know how to complete each of the necessary techniques separately but can you explain how to use them together to produce a product safely?

A number of different skills are needed to carry out the production of a sample of a chemical. This topic looks at the skills of selecting techniques, apparatus and materials and managing safety.

❗ These pages are designed to help you think about aspects of the investigation rather than to guide you through it step by step.

Carrying out procedures

The making of magnesium sulfate from magnesium **carbonate** can be done in several stages using different pieces of apparatus.

Figure 4.32 has some stages of an experimental plan depicted but not all the stages.

In the diagram, the quantities of two chemicals have been measured to make the salt.

DID YOU KNOW?

Most salts have a very high melting point, so a special heater would be needed.

1. Identify the two substances needed to make magnesium sulfate.

2. State how the quantities are measured.

3. State the units of measurement for each substance.

4. Describe the safety measures that need to be taken when adding substances.

5. After the two substances are mixed and stirred as in Figure 4.32b, explain why there is some solid left in the bottom of the solution.

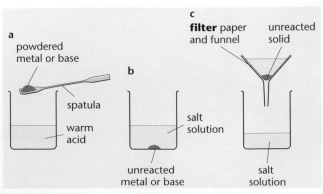

Figure 4.32

6 The excess solid is removed as in Figure 4.32c. Name the process of removal of the solid.

Using apparatus and techniques

The solution is collected after the removal of the excess solid.

To obtain the product from the solution two more processes are needed:

- evaporation
- **crystallisation**.

To make a better product a pure sample is needed. The crystals formed will need to be:

- washed and dried
- recrystallised.

Think about these further questions for making pure magnesium sulfate.

7 Describe how the first sample of dry impure salt is obtained from the solution.

8 Explain how the purity of the salt can be improved.

9 Explain how the purity of the final sample is tested.

Measuring accurately

When a sample of product has been made, it needs to be tested for purity. Sam and Alex made samples of calcium nitrate crystals. They looked up the melting point for these crystals and found it was 42.7 °C. These are their results. Who made the purest sample?

Sam	Mp °C	40.3	40.5
Alex	Mp °C	41.6	41.4

10 Explain the affect of impurities on the melting point.

11 Suggest why it might have been advisable for Sam and Alex to take another temperature measurement.

12 Give the equation for the reaction between magnesium carbonate, $MgCO_3$, and the acid.

13 Calculate the mass of the two substances needed to produce 6 g of $MgSO_4$.

14 Explain why the solid $MgCO_3$ needs to be added *in excess* of the amount calculated.

15 What may happen to the yield if the aim is to increase purity?

16 The uncertainty in reading the thermometer was plus or minus 0.2 °C. Discuss whether the difference in melting point between the data book value and the measurements is significant. Explain which of the results are most accurate.

REMEMBER!

It is helpful to imagine setting up the apparatus needed and draw each part of the procedure by stages.

pH and neutralisation

Learning objectives:

- describe the use of universal indicator to measure pH
- use the pH scale to identify acidic or alkaline solutions
- investigate pH changes when a strong acid neutralises a strong alkali.

KEY WORDS

hydroxide ions
neutralisation
pH
universal
 indicator

We use acids in our normal lives every day. Stomach acid is essential for digesting food. The acidic food we eat is sometimes delicious and sometimes sour. Why are citric acid and ethanoic acid available in supermarkets but hydrochloric acid in the laboratory needs to be used with care and safety glasses need to be worn?

Acids and alkalis

Acids are substances that produce hydrogen ions in aqueous solution.

Examples of acids are:

hydrochloric acid	HCl
nitric acid	HNO_3
sulfuric acid	H_2SO_4
ethanoic acid	CH_3COOH
citric acid	$C_6H_8O_7$

Examples of alkalis are:

sodium hydroxide	NaOH
potassium hydroxide	KOH
ammonia	$NH_3(aq)$

The hydrogen ions they produce have the symbol H^+. They are ions with a positive charge.	Alkalis are substances that make hydroxide ions in aqueous solution. These hydroxide ions have the symbol OH^-. They are ions with a negative charge.

1 Identify the ion in sulfuric acid that makes it acidic.

2 Explain the difference between an acid and an alkali.

The pH scale

When **universal indicator** (UI) is added to solutions it changes colour.

This is because UI is a different colour when the number of hydrogen ions in the solution changes.

The number of hydrogen ions in a solution is related to a scale called the pH scale.

Type of solution	Colour of UI	pH
acidic	red/orange/yellow	0–6
neutral	green	7
alkaline	blue/purple	8–14

Universal indicator can be used to estimate the **pH** of a solution, by matching to a colour/pH chart.

Figure 4.33 Which of these contains sulfuric acid, citric acid, nitric acid or ethanoic acid?

Figure 4.34 Universal indicator colour changes in strong acid, weak acid, neutral, weak alkali and strong alkali

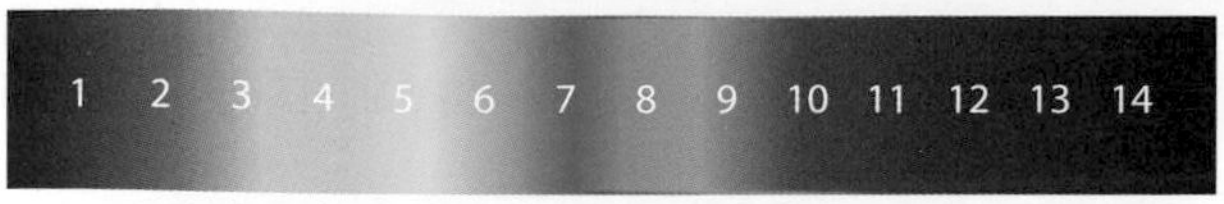

1 2 3 4 5 6 7 8 9 10 11 12 13 14

1 = very acidic, 7 = neutral, 14 = very alkaline

Figure 4.35 pH colour match chart

A more accurate way of measuring pH is to use a pH probe.

3 Universal indicator solution was added to HCl in a conical flask. It was then exactly neutralised by NaOH. Excess NaOH was then added. Describe the colour change and estimate the pH at each stage.

4 What would be the pH of a solution if UI turned green?

Neutralisation

If an acid is added to an alkali, **neutralisation** takes place.

An acid solution has a low pH.
If an alkali is added slowly to an acid, the pH number of the acid will gradually increase.
When it gets to pH 7 the acid is neutralised.

An alkaline solution has a high pH. If acid is slowly added to an alkali, the pH number will gradually decrease.
When it gets to pH 7 the alkali has been neutralised.

The higher the concentration of H^+ ions the lower the pH.

Alkalis contain OH^- ions (**hydroxide ions**).

Neutralisation involves this reaction:

$$H^+_{(aq)} + OH^-_{(aq)} \rightarrow H_2O_{(l)}$$

Neutralisation leaves no free H^+ ions.

The neutralisation reaction takes place with all common acids and common alkalis.

Hydrochloric acid	$H^+ \rightarrow$	Cl^-
Nitric acid	$H^+ \rightarrow$	NO_3^-
Sulfuric acid	$H^+ \rightarrow$	SO_4^{2-}
Ethanoic acid	$H^+ \rightarrow$	CH_3COO^-

Sodium hydroxide	Na^+	$OH^- \rightarrow$
Potassium hydroxide	K^+	$OH^- \rightarrow$
Calcium hydroxide	Ca^{2+}	$OH^- \rightarrow$

$$H^+_{(aq)} + OH^-_{(aq)} \rightarrow H_2O_{(l)}$$

5 Phosphoric acid has the formula H_3PO_4.

a Identify the ions in phosphoric acid.

b Compare the concentrations of H^+ and OH^- ions in phosphoric acid and sodium hydroxide.

HIGHER TIER ONLY

6 Write an *ionic* equation for the reaction between hydrochloric acid and potassium hydroxide.

7 Indigestion is caused by the overproduction of hydrochloric acid in the stomach. Indigestion remedies often contain magnesium hydroxide. This forms a weak alkali in suspension. (see section 4.10)

a Estimate the pH of an indigestion remedy and explain in terms of the ions present.

b Explain how the indigestion remedy works in terms of the ions present.

DID YOU KNOW?

Some indicators show a sudden colour change at one pH value. Universal indicator shows a gradual range of colour changes, as it contains a mixture of different indicators.

burette containing
base of known
concentration
conical
flask
acid of unknown
concentration
containing a few
drops of universal
indicator

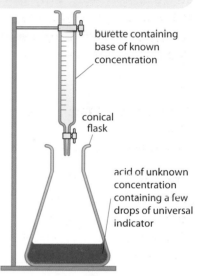

Figure 4.36 Investigating the changes of pH when a base is added to an acid

DID YOU KNOW?

You can plot a graph of the pH change against every 1 cm^3 of alkali you add to acid. You will see a surprising curve from your results.

KEY INFORMATION

Remember: you know that the process of neutralisation creates salts.

acid + alkali → salt + water

Strong and weak acids

Learning objectives:

- explain weak and strong acids by the degree of ionisation
- describe neutralisation by the effect on hydrogen ions and pH
- explain dilute and concentrated as amounts of substance.

KEY WORDS

concentration
dilution
weak
strong

Sulfuric acid, a strong acid, is used in car batteries. Vinegar and citric acid are weak acids that are used in cooking. The terms 'strong' and 'weak' are not the same as 'concentrated' and 'dilute'. You will need to consider the number of particles and the amount they ionise.

HIGHER TIER ONLY

Ionisation

There are two ways to describe any solution of an acid. It can be either concentrated or **dilute** and it can also be either **strong** or **weak**. These do not mean the same thing (they are not synonyms).

Examples of strong and weak acids are:

strong acids	weak acids
hydrochloric	ethanoic
nitric	citric
sulfuric	carbonic

Strong and weak acids both have the same chemical reactions.

Acid reactions are caused by hydrogen ions, H^+.

Strong and weak acids react with:

- metals to make hydrogen
  $Mg + 2H^+ \rightarrow Mg^{2+} + H_2$
- metal carbonates to make carbon dioxide
  $CaCO_3 + 2H^+ \rightarrow Ca^{2+} + H_2O + CO_2$

So why are some acids (H^+A^-) strong and some weak?

Strong:	Weak:
In water all of the acid molecules, HA, become ions (H^+ and A^-).	In water only a few of the acid molecules, HA, become ions (H^+ and A^-), most stay as molecules.
Strong acids ionise completely in water.	Weak acids do not ionise fully. The equilibrium lies to the left.
$HCl \rightarrow H^+ + Cl^-$	$CH_3COOH \rightleftharpoons H^+ + CH_3COO^-$
A high concentration of H^+ means that the pH is low.	A low concentration of H^+ means that the pH is higher.

1   Name the ions produced when sulfuric acid ionises.

2   Explain why citric acid, $H^+(citrate)^-$ is a weak acid.

It is possible to have an acid that is:

strong and concentrated

strong and dilute

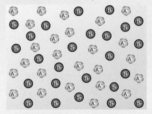

weak and concentrated

weak and dilute

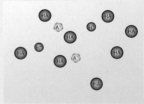

Figure 4.37

pH, neutralisation and titration curves

The pH scale is related to the concentration of H^+ ions.

Strong acids have a lower pH than weak acids.

low pH number = high concentration of H^+	higher pH = lower concentration of H^+ number

For a given **concentration** of aqueous solutions, the stronger an acid, the lower the pH. As the pH decreases by one unit, the hydrogen ion concentration of the solution increases by a factor of 10. This is because pH is based on a logarithmic scale.

During neutralisation of an alkali, if acid is added to the solution the pH will decrease. This can be seen in a titration curve.

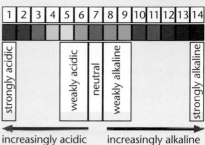

Figure 4.38 The pH scale

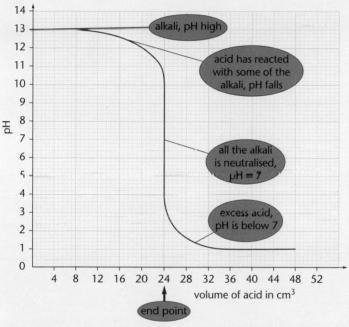

Figure 4.39 Titration curve of a strong alkali (pH 13) neutralised by a strong acid

DID YOU KNOW?

Titration curves for weak acids and weak alkalis do not give a very clear neutralisation point.

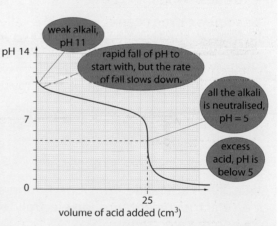

Figure 4.40 Titration curve of a weak alkali (pH 11) neutralised by a strong acid

3 **Suggest a sketch of a titration curve for a weak alkali being added to a strong acid. Take the end point as 25 cm³.**

Concentration

Acid concentration is not the same as H^+ ion concentration. A concentrated solution of a weak acid still has a low concentration of H^+ ions.

Concentration is measured in mol/dm³ or g/dm³.

1 dm³ is the same as 1000 cm³.

100 cm³ is one-tenth of 1000 cm³.

The mass of a solute added to 100 cm³ of water needs to be one-tenth of the mass that would be added to 1000 cm³ of water to be the *same concentration*.

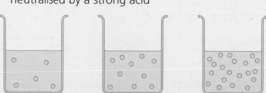

Figure 4.41 Solutions of different concentration

KEY INFORMATION

The concentration of an acid tells us how many moles of acid there are in 1 dm³.

The strength of an acid tells us how much an acid ionises.

4 **A solution is made from 6 g of citric acid added to 100 cm³ of water. What is the concentration of the solution in g/dm³?**

The process of electrolysis

Learning objectives:

- identify reactions at electrodes during electrolysis
- explain why a mixture is used and the anode needs constant replacement
- write and balance half equations for the electrode reactions.

KEY WORDS

electrode
electrolysis
electrolyte
ion migration

Electrolysis can be used to obtain useful products such as oxygen and hydrogen from sulfuric acid, and copper or aluminium from their ores.

Electrolytes

Electrolysis is the process of passing direct current (d.c.) through a solution or melted ionic compound to move the ions apart and so break the compound down and discharge some of the elements at the **electrodes**.

The solution or molten compound is called the electrolyte. The ions need to be free to move about within the solution or liquid to be an **electrolyte**. An electrolyte is a liquid that conducts electricity and is decomposed during the electrolysis.

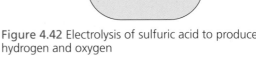

Figure 4.42 Electrolysis of sulfuric acid to produce hydrogen and oxygen

Two *electrodes* are used. They are called the *cathode* and the *anode*. These 'dip into' the solution or 'melt'.

The cathode is the negative electrode.	The anode is the positive electrode.

Electrolytes are made from *ions*. During electrolysis the passing of an electric current through electrolytes causes the ions to move to the electrodes:

positive ions are attracted to the cathode and are called *cations*	negative ions are attracted to the anode and are called *anions*

DID YOU KNOW?

A molten ionic compound involved in electrolysis can be referred to as a 'melt'.

1 Explain why electrolytes need to be molten or in solution.

Positive ions and negative ions

Ions that have a formula with a positive charge, such as H^+ and Na^+ are called cations.

Ions that have a formula with a negative charge, such as OH^- and Cl^- are called anions.

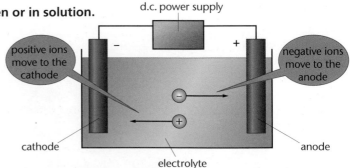

Figure 4.43 The process of electrolysis

Positive ions (cations)		Negative ions (anions)	
H^+	hydrogen	OH^-	hydroxide
Na^+	sodium	Cl^-	chloride
Cu^{2+}	copper	SO_4^{2-}	sulfate
Al^{3+}	aluminium	O^{2-}	oxide

Ions are discharged at the electrodes, producing elements. Positive ions move towards the negative electrode. Negative ions move towards the positive electrode. Opposite charges attract.

KEY INFORMATION

Check the charges on the ions when explaining the migration of the ions to the electrodes.

2 **Identify the two products that would form if molten copper chloride was electrolysed.**

3 **Why do aluminium ions move towards the cathode?**

HIGHER TIER ONLY

Ions discharging

All the **ions migrate** towards the two electrodes.

All the positive ions are attracted to the negatively charged electrode.

All the negative ions are attracted to the positively charged electrode.

When the negative ions reach the anode (the positively charged electrode) they lose electrons to the electrode.

When the positive ions reach the cathode (the negatively charged electrode) they gain electrons at the electrode.

These reactions can be written as half equations.

The half equation for the discharge of the sodium ion would be:

$$Na^+ + e^- \rightarrow Na$$

The half equation for the discharge of the chloride ion would be:

$$Cl^- - e^- \rightarrow Cl$$

However, chlorine exists as molecules of two atoms so two ions need to be discharged.

$$2Cl^- - 2e^- \rightarrow Cl_2$$

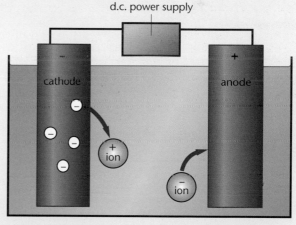

Figure 4.44 Ions discharging at the cathode and anode

DID YOU KNOW?

The discharge of chlorine at the anode is more correctly written as:
$$2Cl^- \rightarrow Cl_2 + 2e^-$$

4 **Write the two half equations for the electrolysis of molten copper chloride.**

5 **Aqueous magnesium bromide solution can be electrolysed.**

 a **Write the half equation for the reaction at the anode.**

 b **Predict the product at the cathode.**

 c **Explain why hydrogen is actually produced at the cathode and where the hydrogen comes from.**

Electrolysis of molten ionic compounds

Learning objectives:

- identify which ions migrate to the cathode and anode
- explain how the ions of a molten electrolyte are discharged
- predict the products of electrolysis of molten binary compounds.

KEY WORDS

anode
cathode
discharged
molten

We have already learned that ions need to be free to move in an electrolyte if electrolysis is to take place. Ions that are not free to move cannot conduct electricity. Discharging the elements from compounds of highly reactive metals can happen if the substances are melted first.

Simple binary electrolytes

A simple binary electrolyte is one that is made up of two ions, for example, lead bromide or copper chloride.

These simple binary electrolytes can conduct electricity if melted. If they are not melted no electricity will flow.

This can be seen by setting up this apparatus in a laboratory, in a fume cupboard.

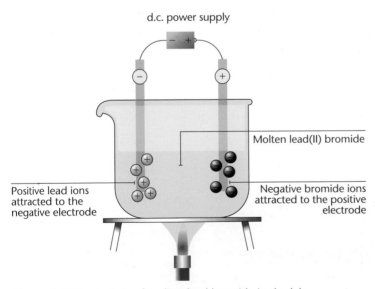

Figure 4.45 Electrolysis of molten lead bromide in the lab

When the lead bromide is cold and solid no electricity will flow and the lamp will not light.

If the crucible is heated and the solid begins to melt a current will flow and the lamp lights.

The ions, because they are now free to move, will migrate towards the electrodes.

1 **Lead makes a positive ion. To which electrode will it move?**

Positive and negative electrodes

When **molten** lead bromide is electrolysed using inert electrodes the ions will migrate towards them.

The lead ion, Pb^{2+}, is positive so will migrate towards the negative electrode.

The negative electrode is the cathode so a lead ion is a cation.

The ion is **discharged** and the metal lead, Pb, is produced at the **cathode**.

The non-metal bromide ion, Br^- is negative so will migrate towards the positive electrode.

The positive electrode is the anode so a bromide ion is an anion.

Two ions are discharged and the non-metal gas bromine, Br_2, is produced at the **anode**.

Another molten binary electrolyte that can be used in the lab is molten copper chloride.

The ions produced are Cu^{2+} and Cl^-.

2 **To which electrode will a chloride ion move?**

3 **Predict the products of the electrolysis of molten sodium bromide at each electrode.**

> **KEY INFORMATION**
>
> The electrodes used are inert. That means they do not react with any element that is discharged during electrolysis. They are normally made of carbon/graphite.

> **DID YOU KNOW?**
>
> Bromine is a brown gas that is a toxic irritant that causes burns. It has an unpleasant smell and bleaching action. That is why you should use a fume cupboard if it is produced.

HIGHER TIER ONLY

Electrode half equations

The ions move towards the electrodes and transfer electrons.

At the cathode	At the anode
Electrons are gained	Electrons are given up

For $PbBr_2$ the half equations are:

$$Pb^{2+} + 2e^- \rightarrow Pb \qquad 2Br^- - 2e^- \rightarrow Br_2$$

Other decompositions can be written as half equations if the formulae of the ions are known.

Electrolyte	Half equation at cathode	Half equation at anode
KCl	$2K^+ + 2e^- \rightarrow 2K$	$2Cl^- - 2e^- \rightarrow Cl_2$
$CuCl_2$	$Cu^{2+} + 2e^- \rightarrow Cu$	$2Cl^- - 2e^- \rightarrow Cl_2$
PbI_2	$Pb^{2+} + 2e^- \rightarrow Pb$	$2I^- - 2e^- \rightarrow I_2$
Al_2O_3	$4Al^{3+} + 12e^- \rightarrow 4Al$	$6O^{2-} - 12e^- \rightarrow 3O_2$

4 **Write the half equations at the cathode and anode for the electrolysis of molten copper bromide.**

Using electrolysis to extract metals

Learning objectives:

- explain the process of the electrolysis of aluminium oxide
- explain why a mixture is used and the anode needs constant replacement
- write half equations for the reactions at the electrodes.

KEY WORDS
....

aluminium
bauxite
carbon anode
cryolite

Aluminium is a very useful metal due to its low density. It is used in alloys for the aircraft industry and in the drinks industry for drinks cans. It is very expensive to manufacture but easy to recycle.

Extracting the more reactive metals

Metals such as iron and zinc can be extracted from their ores using reduction by carbon. However, some metals are too reactive to be extracted by reduction with carbon, or the metal reacts with carbon. These metals are found towards the top of the reactivity series.

These metals can be extracted from molten compounds using electrolysis.

Large amounts of energy are used in the extraction process to melt the compounds and to produce the electrical current for electrolysis. As a result, these metals can be very expensive.

Aluminium is one of the metals extracted in this way. Although it is very abundant in the Earth's crust it is very expensive to extract. This is one of the reasons why there are so many recycling schemes to encourage us to recycle aluminium cans.

Figure 4.46 Recycling aluminium

1 **Explain why sodium has to be extracted by electrolysis rather than by reduction with carbon.**

Manufacturing aluminium by electrolysis

Aluminium is manufactured by the electrolysis of its ore, **bauxite**.

The ore bauxite is a form of aluminium oxide. As the purified ore melts at a very high temperature it is mixed with **cryolite**, as the mixture melts at a lower temperature. It is melted in a steel electrolysis cell.

The molten mixture of aluminium oxide and cryolite is electrolysed in a cell that is a carbon-lined cathode with suspended **carbon anodes**.

Molten aluminium oxide forms positive aluminium ions. The positive ions migrate towards the negative electrode (cathode) and are discharged at the electrode making a molten metal.

The oxide ions are negative ions which migrate to the positive electrode (anode). The oxide ions are discharged as oxygen molecules. The positive electrode (anode) is made of carbon. The anodes, which are suspended in the melt, react with the oxygen evolved to produce carbon dioxide. The carbon anodes are continually reacting and so must be continually replaced.

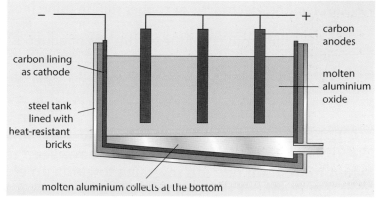

Figure 4.47 Manufacturing aluminium by electrolysis

2 Describe what happens at the two electrodes during the electrolysis of aluminium oxide to make aluminium.

3 The cost of making aluminium is very high. Suggest a reason, other than the cost of materials, why it is so expensive.

HIGHER TIER ONLY

Electrode half equations

$2Al_2O_3 \rightarrow 4Al + 3O_2$

Aluminium ions are positive, Al^{3+}, so they migrate to the negative cathode. Here they gain electrons.

$4Al^{3+} + 12e^- \rightarrow 4Al$

Or per ion:

$Al^{3+} + 3e^- \rightarrow Al$

Oxide ions are negative O^{2-}, so they migrate to the positive anode. Here they transfer electrons to the anode.

$6O^{2-} - 12e^- \rightarrow 3O_2$

As the anodes are made of carbon operating at high temperature they react with the oxygen being evolved to make carbon dioxide.

$3C + 3O_2 \rightarrow 3CO_2$

4 Describe the flow of electrons as aluminium oxide is electrolysed.

5 Write two half equations for extraction of another reactive metal, calcium, from calcium chloride $CaCl_2$ by electrolysis.

6 Write the overall equation and the two half equations for the electrolysis of molten potassium fluoride.

KEY INFORMATION

Aluminium makes a 3^+ ion and oxide a 2^- ion, so the compound is Al_2O_3

Electrolysis of aqueous solutions

Learning objectives:

- explain the electrolysis of copper sulfate using inert electrodes
- predict the products of the electrolysis of aqueous solutions
- represent reactions at electrodes by half equations.

KEY WORDS

electrode
reaction
half equation
preferential discharge

We have already looked at the electrolysis of binary electrolytes in aqueous solution. We have seen that when more than one ion migrates towards an electrode, that it is the most reactive that stays in solution and the *least reactive* that is discharged. The discharge of the least reactive ion is called preferential discharge.

DID YOU KNOW?

There is a list of 'ease of discharge' for negative ions as well as positive ions. This is SO_4^{2-} NO_3^- Cl^- Br^- I^- OH^-

Preferential discharge of ions

The **preferential discharge** is based on the reactivity series.

The relatively least reactive ions are preferentially discharged as elements.

At the cathode	At the anode
H^+ is discharged in preference to Zn^{2+} Al^{3+} Mg^{2+} Ca^{2+} Na^+ and K^+	OH^- is discharged in preference to SO_4^{2-}
Cu^{2+} and Ag^+ are discharged in preference to H^+ (as Cu^{2+} and Ag^+ are less reactive than H^+)	Cl^- and Br^- are discharged in preference to OH^-
H^+ produces hydrogen gas	OH^- produces oxygen gas

1 **Explain whether silver or hydrogen would be discharged first during the electrolysis of aqueous silver nitrate.**

The electrolysis of copper sulfate

The electrolysis of copper(II) sulfate using inert carbon electrodes follows the rules of preferential discharge. Copper is less reactive than hydrogen. What we *see* is that this electrolyte is a blue solution. As the electrolysis happens the blue colour of the solution fades.

We see the cathode being coated with copper and bubbles at the anode.

The copper ions (+) from the solution migrate to the cathode (–) and can be seen to deposit. Let's look at the four ions involved, Cu^{2+}, H^+, SO_4^{2-} and OH^-.

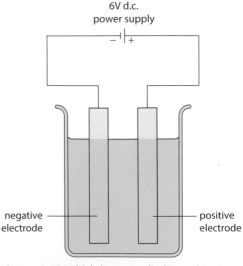

6V d.c. power supply

negative electrode

positive electrode

Figure 4.48 Which ions are discharged in the electrolysis of copper sulfate solution, $CuSO_{4(aq)}$, with inert electrodes?

To the cathode	To the anode
Copper ions (Cu^{2+}) Hydrogen ions (H^+) (from the water)	Sulfate ions ($SO_4{}^{2-}$) Hydroxide ions (OH^-) (from the water)
Hydrogen ions stay in the electrolyte. Copper ions are discharged *in preference* and make copper metal.	Sulfate ions stay in the electrolyte as hydroxide ions are more easily discharged. Hydroxide ions discharge *in preference* and form oxygen gas.

So, if $CuSO_4$ solution is electrolysed with carbon electrodes, copper is formed at the cathode and oxygen is formed at the anode.

2 Sodium chloride was added to a solution of copper(II) sulfate and the mixture electrolysed. Compare the discharge products to the electrolysis of pure copper(II) sulfate solution. Explain your answer.

HIGHER TIER ONLY

Oxidation, reduction and half equations

During electrolysis, positively charged hydrogen ions migrate to the cathode and gain electrons.

$$2H^+ + 2e^- \rightarrow H_2$$

As the ions gain electrons the reactions are *reductions*.

Negatively charged ions migrate to the anode and lose electrons.

As these reactions lose electrons they are *oxidations*.

> **KEY INFORMATION**
> ....
> **O**xidation **I**s electron **L**oss
> **R**eduction **I**s electron **G**ain:
> **OILRIG**

The hydroxide ions migrate to the anode but the **half equation** is not as simple as that for H^+ (where only two electrons were involved). This time four electrons are needed.

$$4OH^- - 4e^- \rightarrow O_2 + 2H_2O$$

(or more correctly $4OH^- \rightarrow O_2 + 2H_2O + 4e^-$)

So for other aqueous solutions the **electrode reactions** are also the same, for example, in the electrolysis of $NaOH(aq)$ or $H_2SO_4(aq)$ they are:

At the cathode: At the anode:

$2H^+ + 2e^- \rightarrow H_2$ $4OH^- - 4e^- \rightarrow 2H_2O + O_2$

In the electrolysis of $NaOH$, H_2 is made and not Na, as Na is much higher up in the reactivity series so H_2 is discharged *in preference*.

However, the electrode reactions in the electrolysis of $CuSO_4(aq)$ with carbon electrodes are:

At the cathode: At the anode:

$Cu^{2+} + 2e^- \rightarrow Cu$ $4OH^- - 4e^- \rightarrow 2H_2O + O_2$

as copper is less reactive than hydrogen and preferentially discharged.

3 Write the half equations of the reactions at the cathode and the anode in the electrolysis of dilute sodium sulfate solution.

4 Write the half equations of the reactions at the cathode and the anode in the electrolysis of dilute copper chloride solution. Explain which is an oxidation and which is a reduction reaction.

5 Predict the half equations at the electrodes for the discharge of phosphoric acid, H_3PO_4, solution.

REQUIRED PRACTICAL

Investigating what happens when aqueous solutions are electrolysed using inert electrodes

KEY WORDS

electrolysis
electrode
inert
electrolyte
cathode
anode

Learning objectives:

- use scientific theories and explanations to develop hypotheses
- plan experiments to make observations and test hypotheses
- apply a knowledge of the apparatus needed for electrolysis including use of inert electrodes and varying electrolytes
- make and record observations.

Using patterns in reactions to form hypotheses about what may happen in other reactions is a very important skill for scientists. Patterns in other types of reactions, such as the reactions of metals with water and acids, may be linked to observations noted in electrolysis. Will you be able to find a link?

❗ These pages are designed to help you think about aspects of the investigation rather than to guide you through it step by step.

Developing hypotheses

You already know the pattern of the reactivity series and that copper can electroplate other metals. Can potassium electroplate other metals?

From observation you already know that copper is less reactive than potassium and that copper from solution will deposit on a graphite **cathode**.

From scientific theory you know that potassium ions tend to form more easily than copper ions.

So one question could be 'does the ability to deposit on an **electrode** link to the reactivity of the metal ion in solution'? From this you could develop more than one hypothesis.

Let's consider one: '*The more reactive the metal, the easier it is to deposit on an electrode*'. Is this reasonable given what you already know or should it be the other way round?

To plan the experiments to test this hypothesis:

What set-up of apparatus and what variables need to be considered when testing this hypothesis?

Figures 4.49 and 4.50 show some ideas.

You might be able to think of some more variables that may need to be kept the same.

DID YOU KNOW?

The reactivity series is closely linked to the electrochemical series.

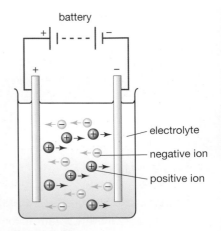

Figure 4.49

Between copper sulfate and potassium sulfate, think about which other five metal sulfate solutions may give a range of results to test the hypothesis.

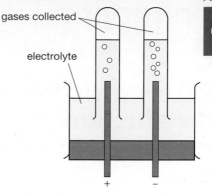

Figure 4.50

1. Identify which **electrode** the metal will deposit on to.

2. Identify which gas may be given off instead of a metal. Describe how you would test for it and what you would see.

Recording and analysing the results

When observations have been gathered, they will need to be interpreted. Sam and Alex tried these **electrolytes** and have started to record these results:

Electrolyte	Prediction for product at cathode	Appearance at cathode or test for gas	Product at cathode	Prediction at cathode supported?
zinc sulfate	zinc	gas popped with lighted spill		
copper sulfate	copper	red/brown metal cover		
sodium chloride	sodium	gas popped with lighted spill		
silver nitrate	silver	silvery/black metal cover		

KEY INFORMATION

A similar set of observations can be made for the reactions at the **anode**.

3. Explain if the observations at the **cathode** support their predictions. Identify the products formed.

4. Explain which pattern they used to make these predictions.

5. Explain why their prediction was correct for copper and silver but incorrect for sodium and zinc. Predict what formed instead of sodium and zinc.

6. Draw a model, using ions, of what happens in the **electrolysis** of copper sulfate solution.

7. Predict the products that form at the **anode**.

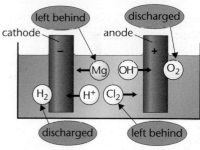

Figure 4.51

Evaluating the experiment

Look at the data gathered by Sam and Alex and reflect on whether the evidence supported the original hypothesis.

8. Explain *the reasons why* these particular products were obtained at each electrode. What is the overall conclusion that can be made from Sam and Alex's results?

9. Explain how your conclusion supports or refutes the hypothesis that the reactivity series is linked to ability of a metal ion in solution to deposit on an **inert** electrode.

KEY CONCEPT

Electron transfer, oxidation and reduction

Learning objectives:

- explain why atoms lose or gain electrons
- explain oxidation and reduction by electron transfer
- relate ease of losing electrons to reactivity.

KEY WORDS

oxidation
reduction
electron loss
electron gain

We have already seen that some elements react more easily than others. We also learned that atoms often need to lose electrons or gain electrons for reactions to take place, and some atoms are able to do this more easily than others. How do all these observations and models relate?

Losing and gaining electrons

Chemical reactions take place when elements rearrange. These elements are made up of atoms.

KEY INFORMATION

Each element has the same *type* of atom but the atoms of each element are different, so all sodium atoms have 11 electrons but all chlorine atoms have 17 electrons.

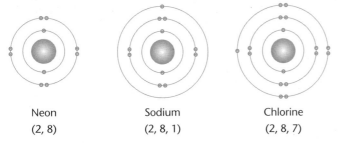

| Neon | Sodium | Chlorine |
| (2, 8) | (2, 8, 1) | (2, 8, 7) |

Figure 4.52 A neon atom, a sodium atom and a chlorine atom

Neon atoms and atoms of other noble gases are unreactive because they have stable electronic configurations. They do not need to gain or lose electrons to achieve stable configurations.

Sodium atoms and all other atoms of *metals* need to *lose* one or more electrons for the metals to react.

Chlorine atoms and atoms of most other *non-metals* need to *gain* one or more electrons to achieve stable configurations (or alternatively *share* electrons).

1 How does an atom of sodium join with an atom of chlorine?

2 Explain why helium and argon are unreactive.

HIGHER TIER ONLY

Oxidation and reduction

When magnesium burns in oxygen it becomes magnesium oxide. This is an **oxidation** reaction.

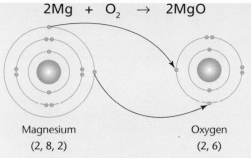

$$2Mg + O_2 \rightarrow 2MgO$$

Magnesium   Oxygen
(2, 8, 2)   (2, 6)

Figure 4.53 Magnesium loses two electrons during the oxidation process.

Oxidation is the **loss of electrons**.

When copper oxide is reacted with hydrogen the oxygen is removed. This is a **reduction** reaction.

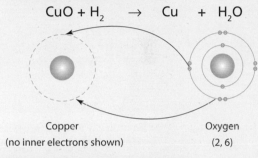

$$CuO + H_2 \rightarrow Cu + H_2O$$

Copper   Oxygen
(no inner electrons shown)   (2, 6)

Figure 4.54 Copper atoms gain back two outer electrons when oxygen is removed.

Reduction is the **gain of electrons**.

These processes also happen at electrodes during *electrolysis*.

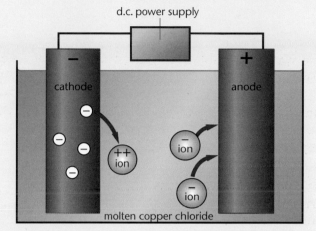

Figure 4.55 Reduction at the negative electrode (cathode), oxidation at the positive electrode (anode)

3 Explain these reactions in terms of oxidation and reduction:
 a zinc to zinc oxide **b** iron(II) oxide to iron

4 Explain these reactions in terms of oxidation and reduction:
 a silver depositing on a cathode
 b oxygen given off at an anode.

Redox reactions and ease of transfer

Some metals lose their electrons more easily than others. It depends on the 'pull' of the nucleus on the electrons leaving. Potassium atoms lose their one electron more easily than magnesium loses its two electrons. A reactivity series can be drawn up to show the order in which metals may lose their electrons more easily.

5 Explain why the electrons from calcium are lost more easily than from magnesium.

6 Explain why sodium is more reactive than magnesium.

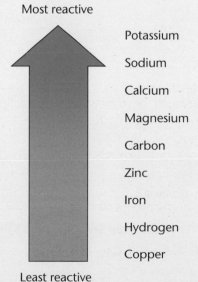

Most reactive

Potassium
Sodium
Calcium
Magnesium
Carbon
Zinc
Iron
Hydrogen
Copper

Least reactive

Figure 4.56 The reactivity series of metals

MATHS SKILLS

Make order of magnitude calculations

Learning objectives:

- use graphs and diagrams to apply the pH scale to acid rain distribution
- calculate the concentration of acids
- calculate the effect of hydrogen ion concentration on the numerical value of pH.

KEY WORDS

logarithmic scale
derived
relative

It is important to be able to calculate the concentration of acids for using them in reactions and monitoring the acidity of rainfall. The concentration of an acid depends on the concentration of hydrogen ions [H+], so we use a **logarithmic scale** to describe acidity more easily, called the pH scale. Can you remember what the numbers of the scale represent?

Acid rain

Non-polluted rainwater has a pH of 5.6. If water has a pH lower than that it is called 'acid rain'. We have seen that the causes of acid rain are polluting gases, such as sulfur dioxide, dissolving in rainwater. Sulfur dioxide comes from burning fossil fuels, especially coal, for electricity generation, transport, heating and industrial processes.

Figure 4.57 shows a map where the rainfall has a pH of less than 5.6 and so is more acidic than usual.

However, there is some better news about sulfur dioxide levels now that power stations have changed. They either burn gas instead, or sulfur dioxide is removed from the emissions in power stations still burning coal. This means that the amount of sulfur dioxide being released has been reducing since 1990 as you can see in the bar chart, Figure 4.58.

DID YOU KNOW?

Robert Angus Smith first showed a relationship between atmospheric pollution and acid rain in 1852 and used the term acid rain in 1872. It took us 100 years to take the issue seriously and do something about it!

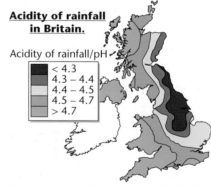

Acidity of rainfall in Britain.

Acidity of rainfall/pH
- < 4.3
- 4.3 – 4.4
- 4.4 – 4.5
- 4.5 – 4.7
- > 4.7

Figure 4.57 Acidity of rainfall in Britain

1. **Which side of the UK experiences rainfall that is the most acidic? Explain how you can tell from the data.**

2. **Estimate the decrease in sulfur dioxide in thousand tonnes equivalent from 1990 to 2006 from the bar chart.**

3. **Calculate the approximate ratio of sulfur dioxide equivalents for 1990 and 2011.**

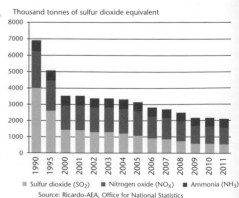

Thousand tonnes of sulfur dioxide equivalent

■ Sulfur dioxide (SO₂) ■ Nitrogen oxide (NOₓ) ■ Ammonia (NH₃)
Source: Ricardo-AEA, Office for National Statistics

Figure 4.58 Acidity of rainfall in Britain

Concentration of acids

The concentration of acids is measured in g/dm^3 or mol/dm^3. The shorthand is to write a square bracket [HA] to mean concentration.

The **relative** *formula mass* (M_r) of nitric acid HNO_3 is calculated from the *relative atomic masses* (A_r)

(A_r: H is 1, N is 14, O is 16) M_r: $HNO_3 = 1 + 14 + (3 \times 16) = 63$

If a solution of nitric acid contains 63 g of acid dissolved in $1\,dm^3$ of water $[HNO_3] = 63\,g/dm^3$. The relative formula mass is also the mass of 1 mole or the *molar mass*, so $[HNO_3] = 1\,mol/dm^3$

$100\,cm^3$ of a solution that contains 6.3 g has decreased both volume and mass by a factor of 10, so the concentration remains the same, $[HNO_3] = 63\,g/dm^3$ and also $[HNO_3] = 1\,mol/dm^3$

$1000\,cm^3$ of a solution that contains 6.3 g has decreased only the mass by a factor of 10 not the volume, so the concentration reduces by a factor of 10, so $[HNO_3] = 6.3\,g/dm^3$ and $[HNO_3] = 0.1\,mol/dm^3$

What does pH mean?

It is a scale that is **derived** from the concentration of hydrogen ions in an acid solution. If the concentration of a strong acid is $0.1\,mol/dm^3$, then the concentration of hydrogen ions, $[H^+]$, is also $0.1\,mol/dm^3$. In standard form this is $10^{-1}\,mol/dm^3$.

The pH number is the 'index' number but positive. So the pH of this solution is pH 1. If the concentration decreases by a factor of 10, the standard form decreases by the index of -1 and the pH increases by positive 1.

Concentration of acid in mol/dm³	Concentration of acid in standard form in mol/dm³	pH
0.1	10^{-1}	1
0.01	10^{-2}	2
0.001	10^{-3}	3
0.0001	10^{-4}	4
0.00001	10^{-5}	5
0.000001	10^{-6}	6
0.0000001	10^{-7}	7

So you can see from the table that as the concentration of the acid decreases by a factor of 10, the pH increases by one unit.

6 Determine the pH of a solution with $[H^+] = 0.000001$.

7 Estimate the concentration of hydrogen ions for a solution that turned universal indicator green.

KEY INFORMATION

You do not have to remember how to use mol/dm^3 when calculating concentrations of acids. It is included here to show you how pH is derived.

4 **9.13 g of hydrochloric acid was dissolved in $250\,cm^3$ of water. Calculate the concentration of hydrochloric acid in g/dm^3.**

5 **0.46 g of methanoic acid, HCOOH, was dissolved in $200\,cm^3$ of water. Calculate the concentration of methanoic acid in g/dm^3.**

DID YOU KNOW?

The pH value is actually a value on a logarithm scale. The pH equals the negative log to the base 10 of the hydrogen ion concentration. This is more elegantly written as: $pH = -\log_{10}[H^+ (aq)]$.

Check your progress

You should be able to:

identify that metals react with oxygen to form metal oxides → explain reduction and oxidation by loss or gain of oxygen → identify metal oxides as bases or alkalis

describe the reactions, if any, of metals with water or dilute acids to place these metals in order of reactivity → explain how the reactivity is related to the tendency of the metal to form its positive ion → deduce an order of reactivity of metals based on experimental results

identify substances oxidised or reduced by gain or loss of oxygen → explain how extraction methods depend on metal reactivity → interpret or evaluate information on specific metal extraction processes

use experimental results of displacement reactions to confirm the reactivity series → use the reactivity series to predict displacement reactions → write ionic equations for displacement reactions

describe how to make salts from metals and acids → write full balanced symbol equations for making salts → use half equations to describe oxidation and reduction

describe how to make pure, dry samples of soluble salts → explain how to name a salt → derive a formula for a salt from its ions

describe the use of universal indicator to measure pH → use the pH scale to identify acidic or alkaline solutions → investigate pH changes when a strong acid neutralises a strong alkali

explain weak and strong acids by the degree of ionisation → describe neutralisation through the effect on hydrogen ions and pH → explain the terms dilute and concentrated as the amounts of substances dissolved

describe the use of an electrolyte, anode and cathode in an electrolysis → identify reactions at electrodes during electrolysis → write half equations for the electrode reactions and complete and balance half equations

explain why electrolytes need to be molten to conduct electricity → describe the products of molten binary electrolytes → predict the products of the electrolysis of binary ionic compounds in the molten state

explain why some metals need to be extracted by electrolysis → explain the process of the electrolysis of aluminium oxide → explain which non-metals are formed at the anode in preference

use apparatus to electrolyse aqueous solutions in the laboratory → explain which metals (or hydrogen) are formed at the cathode in preference → predict the products of the electrolysis of aqueous solutions containing a single ionic compound

explain the electrolysis of copper sulfate using inert electrodes → predict the products of the electrolysis of aqueous solutions → represent reactions at electrodes by half equations

Worked example

Kim and Jo are electrolysing dilute sulfuric acid.

1 **Identify the substance seen at the anode.**

   a nitrogen   **b** hydrogen   **c** sulfur   (**d oxygen**)

> The answer oxygen is correct.

2 **Describe how they will test for hydrogen gas.**

   It pops with a lighted splint.

> This test is correct.

3 **Construct the half equation for the discharge of hydrogen at the electrode.**

   $H^+ + e^- \rightarrow H_2$

> The charge on the ion and the gain of an electron are correct. The molecule H_2 is correct. The equation needs to be balanced,
> $2H^+ + 2e^- \rightarrow H_2$

4 **Next Kim and Jo want to electrolyse copper sulfate. Jo says that they cannot use solid copper sulfate. Explain why.**

   The ions need to be free to move.

> This answer is partly correct. The ions need be free to move to conduct electricity, so need to be molten or in solution.

5 **They choose to use a solution. They pass a current through the solution of copper sulfate, using carbon electrodes.**

   a **Describe what will happen at the cathode.**

     It will get a coat of pink/brown copper.

> The answer is correct.

   b **Describe what they will see at the anode.**

     It will disintegrate.

> The student needs to be clear that this happens with copper electrodes, not carbon, and is used in the purification process. The answer should be that bubbles of oxygen will appear.

   c **What will they see happening to the copper sulfate solution?**

     It will stay blue.

> The answer is again confused with the purification process. The correct answer is that the blue colour disappears. This is due to the copper depositing from the solution on to the electrode.

6 **Construct the half equation for the discharge of copper at an electrode and explain whether this reaction is oxidation or reduction. Explain why copper is deposited and hydrogen is not evolved.**

   $Cu^{2+} + 2e^- \rightarrow Cu$
   This is reduction as electrons are gained (RIG).
   Copper is deposited because it is more reactive than hydrogen.

> The half equation is correct. Reduction is correct – no need for the memory aid. Copper is less reactive than hydrogen (which is why it is deposited).

End of chapter questions

Getting started

1. Which oxide is a base?

   1 Mark

   a carbon dioxide   **b** magnesium oxide   **c** sulfur dioxide   **d** silicon dioxide

2. Which metal is most reactive?

   1 Mark

   a sodium   **b** iron   **c** zinc   **d** magnesium

3. Write down two substances needed to make iron metal from its ore.

   2 Marks

4. Zinc is reacted with sulfuric acid. Bubbles form. What gas is this?

   1 Marks

   a oxygen   **b** carbon dioxide   **c** hydrogen   **d** chlorine

5. 1 g of metal carbonate dissolves completely in acid, but 2 g does not dissolve. How many grams do you add to the same volume of acid to make a salt?

   1 Mark

6. Sodium carbonate is added to a flask containing hydrochloric acid and a few drops of universal indicator. A small excess of sodium carbonate was added. Explain what happens to the colour in the flask.

   2 Marks

7. The diagram below shows four different metals reacting with sulfuric acid solution.

   2 Marks

   a Determine the order of **decreasing** reactivity for metals **D** to **G**.

   b Explain what makes one metal more reactive than another.

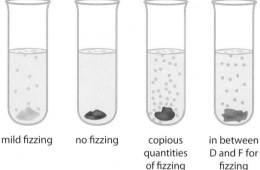

mild fizzing   no fizzing   copious quantities of fizzing   in between D and F for fizzing

Going further

8. Estimate the pH of hydrochloric acid solution.

   1 Mark

9. Which metal oxide can be reacted with carbon to produce the metal?

   1 Mark

   a aluminium oxide   **b** sodium oxide   **c** iron(III) oxide   **d** magnesium oxide

10. Explain why solid lead bromide cannot be electrolysed.

    2 Marks

11. Describe the stages of making pure zinc sulfate salt crystals from sulfuric acid. Give equations where appropriate.

    4 Marks

12. Some experiments were carried out on metals and solutions of metal salts.

    - Zinc was added to copper(II) sulfate solution. Copper was deposited.

    - Metal X was added to zinc sulfate solution. Zinc was not deposited.

    - Metal X was added to copper(II) sulfate solution. Copper was deposited.

    Place the metals in order of reactivity. Justify your answer.

    2 Marks

More challenging

13 Explain oxidation in terms of electron transfer using sodium chloride, NaCl, as an example.

14 Methanoic acid, HCOOH, is a weak acid. Define the term weak acid.

1 Mark

15 Aqueous potassium hydroxide is added to a solution of phosphoric acid, H_3PO_4. State the type of reaction and write the balanced ionic equation including state symbols.

2 Marks

16 Silver sulfate solution was electrolysed. Predict the products at the electrodes.

2 Marks

17 In an experiment a student reacted metals with solutions of sulfates of the other metals. One metal, X, was unknown. These are the results:

4 Marks

	Mg	Fe	Cu	X
$MgSO_4$		✗	✗	✗
$FeSO_4$	✓		✗	✓
$CuSO_4$	✓	✓		✓
XSO_4	✓	✗	✗	

a Describe where **X** lies in the reactivity series, giving reasons.

b For one of the reactions, give a balanced chemical equation.

Most demanding

18 Solid magnesium carbonate, $MgCO_3$, was added to an aqueous solution of nitric acid in a conical flask.

a State **two** observations that could be made.

b Write a balanced chemical equation for the reaction, including state symbols.

2 Marks

19 Tin is present as tin(IV) oxide, SnO_2, in cassiterite ore. The ore is heated with carbon and other materials such as limestone and sand.

a Write a balanced equation for the reaction of tin(IV) oxide with carbon.

b Explain what is happening in terms of oxidation and reduction.

c Part of the reactivity series is as follows: K, Na, Mg, Al, C, Zn, Fe, Sn, Pb, H_2, Cu, Ag, Au. Explain why it is not necessary to use electrolysis to extract tin.

4 Marks

20 An aqueous solution of calcium iodide was electrolysed.

a Predict the products at the electrodes.

b Write the half equations occurring at the electrodes.

4 Marks

Total: 40 Marks

ENERGY CHANGES

IDEAS YOU HAVE MET BEFORE:

ENERGY IN REACTIONS

- Fuels burn to give out energy.
- Some chemicals break down on heating.
- Some chemical reactions release energy.

FOLLOWING ENERGY CHANGES

- Chemical energy can transfer to electrical energy.
- Chemical energy can transfer to light energy and heat energy.
- Energy in joules is written on food packets.

EXPLAINING ENERGY CHANGES

- A temperature rise means that energy is being transferred out.
- Energy is transferred when something is moving.
- Energy can be stored.

THE FIRE TRIANGLE

- Fuel needs oxygen and an energy source to start it burning.
- Water on a burning wood cools it down to stop the fire.
- The fire triangle shows that three things are needed for burning.

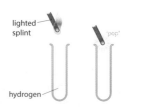

lighted splint

'pop'

hydrogen

MEASURING TEMPERATURES

- The boiling point of water is 100 °C.
- Thermometers work by liquid expansion next to a scale.
- Thermometers for body temperature need a small range scale.

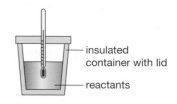

insulated container with lid

reactants

IN THIS CHAPTER YOU WILL FIND OUT ABOUT:

WHAT ARE THE ENERGY CHANGES IN REACTIONS?

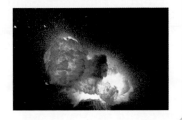

- Exothermic reactions are those that give out energy.
- Endothermic reactions cause a drop in temperature.
- Energy transferred can be calculated from bond energies.

HOW DO WE REPRESENT ENERGY CHANGES?

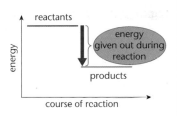

- Energy changes are represented by energy profile diagrams.
- Activation energy needs to be overcome for a reaction to occur.
- The energy level of the products depends on the type of reaction.

HOW CAN WE EXPLAIN ENERGY CHANGES?

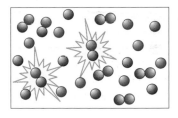

- Collision theory explains that particles are moving and colliding.
- Reactions take place if successful collisions take place.
- Energy changes can be calculated using bond energies.

WHAT DOES THE PROFILE OF BURNING FUEL LOOK LIKE?

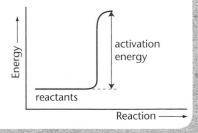

- The energy of the fuel and oxygen reactants must exceed the activation energy.
- Burning fuels is an exothermic reaction.
- In exothermic reactions, product energy is lower than reactant energy.

HOW CAN ENERGY CHANGES BE CALCULATED?

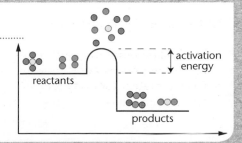

- Each bond has a calculated bond energy.
- Bond breaking is an endothermic process.
- Bond making is an exothermic process.

KEY CONCEPT

Endothermic and exothermic reactions

Learning objectives:

- identify exothermic and endothermic reactions from temperature changes
- evaluate the energy transfer of a fuel
- investigate the variables that affect temperature changes in reacting solutions.

KEY WORDS

endothermic
energy transfer
exothermic
surroundings

How could you find out which fuel gives out most heat energy? How do you choose which heat pack to use to warm yourself? How do you choose which cool pack to use to treat an injury strain? To answer these questions, you need to investigate reactions and temperature changes.

Exothermic and endothermic reactions

Chemical reactions happen when reactants change into products.

Chemical reactions can either *give out* or *take in* energy.

Reactions that *give out* energy to the surroundings (release energy) are exothermic.	Reactions that *take in* energy from the surroundings (absorb energy) are endothermic.
Measuring temperature changes shows the type of reaction:	
if the temperature of the surroundings *goes up*, it is an *exothermic* reaction. ↑	if the temperature of the surroundings *drops*, it is an *endothermic* reaction. ↓
Exothermic reactions include combustion, many oxidation reactions and neutralisation.	Endothermic reactions include thermal decompositions, the reaction of citric acid and sodium hydrogencarbonate, and photosynthesis.
Everyday uses of exothermic reactions include self-heating cans and hand warmers.	Some sports injury packs, which you use to cool a sprain, are based on endothermic reactions.

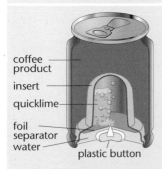

coffee product
insert
quicklime
foil separator
water
plastic button

Figure 5.1 This uses an exothermic reaction.

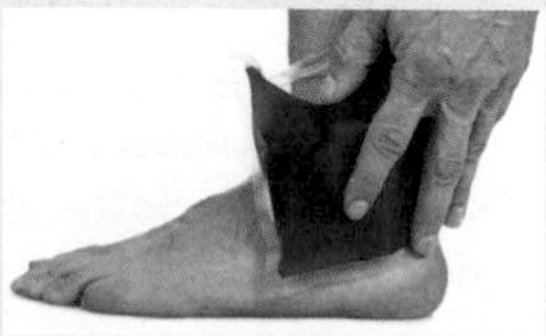

Figure 5.2 This uses an endothermic reaction.

1 Hydrochloric acid was added to sodium hydroxide and the temperature change was measured. The temperature increased by 5 °C. Explain whether this was an exothermic or an endothermic reaction.

2 Explain why photosynthesis is an endothermic reaction.

Mending injuries

Sports injury packs rely on **endothermic** reactions.

	Mass of citric acid (g) added to 10 g of sodium hydrogencarbonate in water				
	6	7	8	9	10
Temperature at start °C	23	22	23	24	21
Temperature at end °C	19	16	15	14	10
Temperature difference °C					

3 Look at the table above. Complete the data. A sports injury pack needs to lower the temperature by 7.5 °C. Estimate the mass of citric acid that needs to be added to 10 g of sodium hydrogencarbonate in order to achieve this.

Investigating and evaluating

Neutralisation is an **exothermic** reaction. A number of variables affect temperature changes of reacting solutions during neutralisations.

These can be investigated using apparatus such as in Figure 5.3.

Phil and Chris took these results when adding 5 cm³ portions of acid to an alkali.

Volume of acid cm³	0	5	10	15	20	25	30	35	40	45
Temperature °C	21	24	26	27	29	31	28	27	25	24

4 Draw a graph of the results from the table above. Explain when neutralisation took place. Suggest what was happening in the reaction at each stage, as indicated by the change in temperature.

5 Explain why a polystyrene cup was used during the experiment.

6 Phil and Chris want to investigate how the temperature change is affected by other factors. Suggest which variables could be changed and which should be kept constant.

7 Identify **one** source of error in the method and **one** possible source of error in the measurements. Suggest modifications to improve accuracy and minimise these errors.

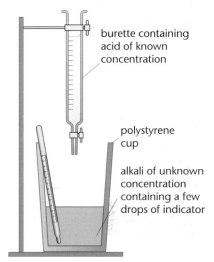
burette containing acid of known concentration

polystyrene cup

alkali of unknown concentration containing a few drops of indicator

Figure 5.3 Measuring temperature rises during neutralisation

REQUIRED PRACTICAL

Investigate the variables that affect temperature changes in reacting solutions, such as acid plus metals, acid plus carbonates, neutralisations, displacement of metals

KEY WORDS

temperature
carbonate
variables

Learning objectives:

- use scientific theories and explanations to develop hypotheses
- plan experiments to make observations and test hypotheses
- evaluate methods to suggest possible improvements and further investigations.

Using previous knowledge to develop a hypothesis for further investigation is an important skill for all scientists. Temperature change is measurable when iron is added to acid. If the acid is changed to another metal, is the effect the same? Some metals are more reactive than others; does this pattern allow us to predict the change?

This topic looks at a few of the skills you need to use to explore the factors that affect a **temperature** change.

Developing a hypothesis

We need to ask a question first, such as 'what are temperature changes affected by'?

Temperature changes may be affected by any of the **variables** that can be changed in a reaction.

Here are four reactions that could be investigated and some of the variables that could be changed:

> ❗ These pages are designed to help you think about aspects of the investigation rather than to guide you through it step by step.

DID YOU KNOW?

Temperature changes during reactions are used in heat packs for pain relief and for heating instant coffee drink packs.

Which variables affect the temperature change when	Variables that could be changed			
Acids react with metals	type of metal	type of acid	concentration of acid	size of pieces of metal
Acids react with **carbonates**	type of carbonate	type of acid	concentration of acid	size of pieces of carbonate
Acids react with alkalis	type of alkali	type of acid	concentration of acid	concentration of alkali
A metal is added to a solution of ions of another metal				

You might be able to think of some more.

Do we have previous knowledge of reactions or patterns that might help us focus down?

1 Suggest whether different reactants and changing concentration would affect the temperature change.

2 Fill in the bottom row for investigating temperature changes in a displacement reaction.

Planning an investigation

What is done in an investigation into any of these reactions is that one of the variables is selected and altered. All the other factors are kept the same.

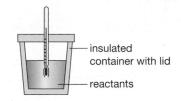

insulated container with lid

reactants

Figure 5.4

3 Explain why it is important to change just one variable at a time.

4 A hypothesis might be: 'The smaller the piece of metal the bigger the temperature change when reacting with acid.' Suggest which variables you would keep the same.

5 Some students developed the hypothesis: 'The more concentrated the acid the greater the temperature change when reacting with carbonates.' They gathered these results:

Concentration of acid in mol/dm³	Temperatures in °C					
	1st trial			2nd trial		
	Start temp	Final temp	Temp change	Start temp	Final temp	Temp change
2.0	20.0	27.3	7.3	21.0	28.2	
1.5	22.0	27.2	5.2	22.4	27.8	
1.0	18.5	23.3	4.8	20.6	31.7	
0.5	21.6	25.5	3.9	21.2	24.9	

a Calculate the temperature changes for the 2nd trial.

b Identify the anomaly in the data. Suggest how the students should deal with this result.

c Suggest what conclusion the students came to with this data.

Evaluating the experiment

When experiments have been completed a reflection should take place. Could the techniques have been improved? How well does the data gathered support the predictions and conclusions made? Were the observations reasonable or could they have another explanation? For example, 'did particle size affect the final temperature change or just the rate at which it achieved the highest temperature?'

6 Explain how the evidence supports or refutes the original hypothesis developed by the students.

7 Explain any problems with the solid carbonates that may affect the repeatability of the experiment.

8 Explain why heat loss needs to be prevented.

> **KEY INFORMATION**
>
> Patterns such as the reactivity series or evidence from rate investigations may be a starting point for developing ideas for hypotheses.

Reaction profiles

Learning objectives:

- draw simple reaction profiles (energy level diagrams)
- use reaction profiles to identify reactions as exothermic or endothermic
- explain the energy needed for a reaction to occur and calculate energy changes.

KEY WORDS

exothermic
product
profile
reactant

All chemical reactions need enough energy to start. Sometimes energy needs to be taken in from the surroundings. We know that petrol burns easily in air. But you can have a beaker of petrol on the bench in your lab without anything happening. The reaction needs some energy to start. A lit match works well.

Colliding particles

Chemicals are made up from 'particles' that have their own energy.

When two chemicals react their particles have energy and are moving.

As the particles meet they collide.

Some collisions do not result in a reaction. They are unsuccessful. Some collisions do result in a reaction. They are *successful.* They have enough energy to react.

These successful colliding particles have at least the minimum energy needed to make a reaction. They have enough energy to activate the reaction.

This energy needed to start is called the *activation energy.*

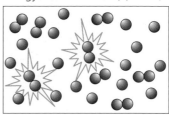

Figure 5.5 The reactants of two reactions. Reactants (a) have lower energy than reactants (b).

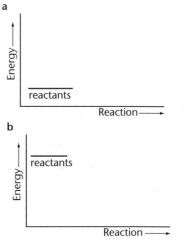

Figure 5.6 Some particles collide without sufficient energy but some particles collide with sufficient energy to react.

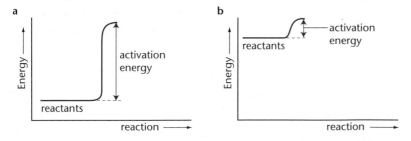

Figure 5.7 Activation energy for the reactions. Reactants (a) have lower energy than reactants (b) and a higher activation energy.

1 **Explain what provides the activation energy in a firework.**

Endothermic and exothermic reactions

Once the activation energy is reached a reaction can proceed until the products are made.

The products will have energy too, once they are made. Their energy level can be drawn onto the diagram of the reaction progress.

DID YOU KNOW?

Energy is conserved in chemical reactions. The amount of energy in the universe at the end of a chemical reaction is the same as before the reaction takes place. If a reaction transfers energy to the surroundings the product molecules must have less energy than the reactants, by the amount transferred.

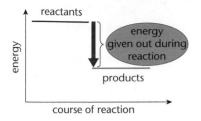

Figure 5.8 A reaction profile for an exothermic reaction

In Figure 5.8 you can see that the product energy level is lower than the reactant energy level. This means that energy has been *given out* (to the surroundings) in the reaction.

This is an *exothermic* reaction

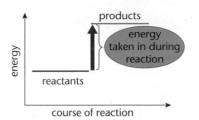

Figure 5.9 A reaction profile for an endothermic reaction

In Figure 5.9 you can see that the product energy level is higher than the reactant energy level. This means that energy has been *taken in* (from the surroundings) in the reaction.

This is an *endothermic* reaction.

Now we have all the stages, we can draw complete reaction **profiles**.

Reaction profiles can be used to show three things:

1 the relative energies of **reactants** and **products**

2 the activation energy

3 the overall energy change of a reaction.

A reaction profile for an **exothermic** reaction can be drawn as shown in Figure 5.10.

2 Draw a complete energy profile for an endothermic reaction.

3 Tom and Nabila reacted zinc with acid. They found that the temperature of the acid increased as the zinc was added. Draw a reaction profile for this reaction.

HIGHER TIER ONLY

Breaking and making bonds

Chemical reactions can be thought of as happening in two stages:

* In stage 1 energy is needed to break the bonds in the reactants into separate atoms. This energy is called activation energy. It is an endothermic process.
* In stage 2 the separated atoms combine to form the products, and energy is released. This is an exothermic process.

If less energy is needed to break bonds than released on making new bonds, then a reaction is exothermic overall.	*If more energy is needed to break bonds than released on making new bonds, then a reaction is endothermic overall.*

4 The reaction between hydrogen and oxygen to form water is exothermic. Explain why, in terms of the bond breaking and making.

REMEMBER!

Chemical reactions can occur only when reacting particles collide with each other and with sufficient energy. The minimum amount of energy that particles must have to react is called the activation energy. A lighted match heats petrol vapour enough to make the particles move with more energy and collide with oxygen particles successfully. The reaction that follows is usually obvious.

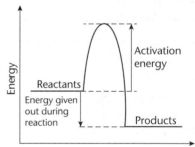

Figure 5.10 A complete reaction profile for an exothermic reaction

DID YOU KNOW?

Catalysts provide a different pathway which has a lower activation energy

Energy change of reactions

Learning objectives:

*   describe the energy changes in bond breaking and bond making
*   explain how a reaction is endothermic or exothermic overall
*   calculate the energy transferred in chemical reactions using bond energies.

The bond energy of bonds between different atoms can be measured. This data is widely available and can be used to calculate the overall energy of reactions. This is an example of using experimental data to provide data that can be applied theoretically to other reactions.

HIGHER TIER ONLY

Breaking and making bonds

Bond *breaking* is an *endothermic* process	Bond *making* is an *exothermic* process

Burning fuels is a clear example where we can see that overall energy is given out. They are exothermic reactions.

However, the overall process is a combination of both **endothermic** and **exothermic** processes.

Bond breaking must take place first, followed by **bond making**.

The bonds of methane and of oxygen are broken. Energy is taken in.	The bonds of carbon dioxide and of water are made. Energy is given out.

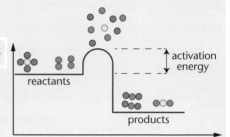
Figure 5.11 Energy is taken in to break bonds. Once the activation energy has been exceeded, energy is given out when making bonds.

In this reaction more energy is given out than taken in so the overall reaction is exothermic.

Figure 5.12 An exothermic reaction

Figure 5.13 Photosynthesis. Is it an endothermic or an exothermic reaction?

1 Draw an energy profile and explain what happens in terms of bond breaking and bond making in this reaction. During the reaction the temperature goes down.

$$Na_2CO_3 + 2CH_3COOH \rightarrow 2CH_3COONa + H_2O + CO_2$$

Bond energies

To decide if a chemical reaction is exothermic or endothermic, the amount of energy needed for bond breaking must be compared with the amount of energy released when new bonds are formed.

REMEMBER!

....

Energy must be supplied to break bonds in the reactants. Energy is released when bonds in the products are formed.

The energy needed to break bonds and the energy released when bonds are *formed* can be calculated from *bond energies*.

The difference between the		
sum of the energy needed to break bonds in the reactants	and	sum of the energy released when bonds in the products are formed
is the overall energy change of the reaction.		

In an endothermic reaction, the energy needed to break existing bonds is greater than the energy released from forming new bonds.	In an exothermic reaction, the energy released from forming new bonds is greater than the energy needed to break existing bonds.

Some common bond energies (in kJ/mol) are:

C—H	C—C	C—O	C=O	O=O	H—O
412	368	352	805	498	465

For example in the reaction:

$$C_3H_8 + 5O_2 \longrightarrow 3CO_2 + 4H_2O$$

the sum of the bond energies (in kJ/mol) are:

Reactants			Products	
$2 \times$ C—C	$= 2 \times 368$	$= 736$	$3 (2 \times$ C=O$) = 3 (2 \times 805) = 4830$	
$8 \times$ C—H	$= 8 \times 412$	$= 3296$	$4 (2 \times$ H—O$) = 4 (2 \times 465) = 3720$	
$5 \times$ O=O	$= 5 \times 498$	$= 2490$		
	$= 6522$		$= 8550$	

The difference is 2028 kJ/mol. This means that the reaction is exothermic (energy released is indicated by the – sign) and 2028 kJ/mol is released.

KEY INFORMATION

It is a convention that an exothermic reaction has a minus (negative) sign before the number and an endothermic reaction has a plus (positive) sign.

2 What is sum of the bond energies for 2 moles of the molecule below?

Calculations

Bond energies can be used to find the overall energy change in a reaction. Use the bond energies in the table to complete the calculations in question 3.

3 Pentane, C_5H_{12}, can be used as a fuel. Calculate the energy change for the reaction:

$$C_5H_{12} + 8O_2 \rightarrow 5CO_2 + 6H_2O$$

MATHS SKILLS

Recognise and use expressions in decimal form

KEY WORDS

decimals
integers
scale
fraction

Learning objectives:

- read scales in integers and using decimals
- calculate the energy change during a reaction
- calculate energy transferred for comparison.

It is important to be able to calculate the energy change during a reaction. If we want to make comparisons between fuels or foodstuffs, for example, we need to know the amount of energy transferred when they burn.

Can you remember the unit for measuring energy changes?

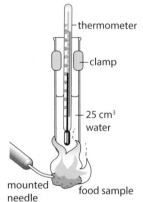

Figure 5.14

Figure 5.15

Temperature changes and energy changes

The easy part of any experiment is to measure the temperature.

We can read a thermometer with a **scale** that goes up in whole numbers (Figure 5.16). Whole numbers are called **integers**.

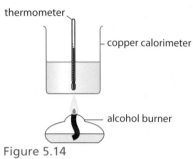

Figure 5.16

Figure 5.17

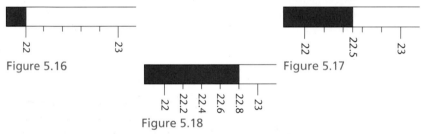

Figure 5.18

We can also read a thermometer with a scale that goes up in half degrees. Between two whole numbers is a 'half degree' mark. Figure 5.17 shows a reading of $22^1/_2$ °C which is also 22.5 °C.

$22^1/_2$ °C is a reading with a **fraction** form and 22.5 °C is a reading with a **decimal** form.

Figure 5.18 shows a thermometer with a scale that increases by 2/10th each time or in decimal form by 0.2 °C.

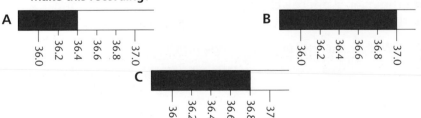

KEY INFORMATION

Energy change is *not* the same as temperature change. Temperature is *measured* but energy is *calculated*. Temperature is measured in °C (degrees Celsius). Energy is calculated in joules (J) or kilojoules (kJ).

1 **A thermometer is used to measure a person's body temperature. The thermometer reads 36.8 °C. Which thermometer was used to make this recording?**

A

B

C

2 Normal body temperature can be 0.4 °C higher than 36.8 °C. What is this higher temperature?

Decimal form

It is possible to write numbers that are smaller than 1 either as fractions or in decimal form.

Decimal form begins when 1 is divided by 10. This is written as 0.1. The number 1 has moved down the column hierarchy to the right of the decimal point.

Can you think what 1 divided by 100 would be? The number 1 would move two places down to be 0.01

Fraction	Decimal	0	.	0	00	000	0000
one-tenth	zero point one	0	.	1			
one-hundredth	zero point zero one	0	.	0	1		
one-thousandth	zero point zero zero one	0	.	0	0	1	
one-ten thousandth	zero point zero zero zero one	0	.	0	0	0	1

Comparing fuels

Alcohols are easily burned in oxygen and can be used as fuels. The amount of energy they transfer out *cannot* be measured, but this energy can be transferred to a body of water. The temperature increase of the water *can* be measured.

Imagine if the energy from some fuel is transferred either to 10 g of water or to 50 g of water. What would happen? The temperature of the 10 g of water would go higher. We can see that our calculation needs to take account of the *mass* of the water.

Now imagine that we transferred the energy to olive oil, not to water. Would the temperature go up by the same amount? Possibly not. So, we also need to take account of the *specific heat capacity* of the receiving liquid.

From these three quantities (temperature increase, mass of the receiving liquid, specific heat capacity of the receiving liquid) we can calculate the energy transferred from an alcohol to water (Figure 5.19) using the equation:

energy transferred by fuel = mass of water × specific heat capacity of water × temperature change of water

energy transferred by fuel = 100 × 4.2 × temperature change of water

$$E = m \times c \times (t_2 - t_1)$$

3 What would be the decimal if 1 was divided by 1000?

4 What would be the decimal if 1 was divided by 100 000?

5 What is 1 divided by to make the number 0.000 000 01?

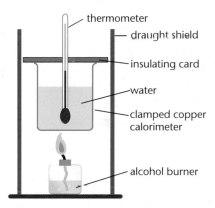

Figure 5.19

6 Propanol burned to raise the temperature of 100 g of water by 24 °C. How much energy did it transfer?

7 When 3.8 g of propanol was burned, the energy transferred was 13 300 J. How many joules were transferred by 1 g of propanol?

DID YOU KNOW?

The specific heat capacity of water is a fixed quantity, which is 4.2 J/g/°C.

Check your progress

You should be able to:

know that energy is conserved	recognise that in a chemical reaction energy can be transferred to the surroundings	recognise that in a chemical reaction energy can be transferred in from the surroundings
identify exothermic and endothermic reactions from temperature changes	identify exothermic reactions as causing a temperature rise	identify endothermic reactions as causing a temperature decrease
identify examples of exothermic reactions	identify examples of endothermic reactions	explain and evaluate the uses of some exothermic and endothermic reactions
investigate changes in temperature of different reactions	investigate the variables that affect temperature changes in reacting solutions	explain how the variables investigated affect temperature changes
draw simple reaction profiles (energy level diagrams)	use reaction profiles to identify reactions as exothermic or endothermic	explain the idea of activation energy and calculate energy changes in a reaction using an energy level profile
calculate the energy change when transferred to 100 g of water using the temperature difference	calculate the energy change using a given equation	calculate energy changes in a reaction given information on mass of reactant
recognise that energy transfer during a reaction is due to bonds being broken and then new bonds being made	describe the energy changes in bond breaking as endothermic and bond making as exothermic and explain how the energy of a reaction is calculated overall	calculate the energy transferred in chemical reactions using bond energies

Worked example

1 When ethanol is burned and heats 100 cm³ water the temperature increases. What type of reaction is this?

(exothermic)   reversible   endothermic

> The answer exothermic is correct.

2 Draw an energy profile to show an endothermic reaction.

> The profile is for an exothermic reaction. The energy level of the products needs to be higher than the level for the reactants. The activation energy is the *difference between* the energy levels of the reactants and products not the initial energy level.

3 Tom investigates temperature changes during a reaction with cold acid. Describe how he could minimise heat losses.

He could put a lid on the beaker.

> This answer is partially correct but he could also insulate the beaker or vessel.

4 What are the type of energy changes that take place for:

a bond breaking   **b** bond making

a) endothermic   b) exothermic

> These two answers are correct.

5 Fill in the missing numbers in the table. Use the equation:

energy = mass of water × specific heat capacity of water
        × temperature change

to calculate the energy change when each fuel transfers energy to 100g of water producing a temperature *change*. The specific heat capacity of water is 4.2 J/g/°C.

> The data on the table is completed correctly.

Number of carbon atoms in alcohol	Initial water temperature, in °C	Final water temperature, in °C	Temperature change, in °C	Energy transferred, in J
3	22	52	30	12 600
4	25	63	38	15 960
5	23	67	44	18 480

6 Write a conclusion you could draw from the investigation in question 5 and what further data you would need to justify it.

The alcohol with the biggest number of carbon atoms produces the most energy. You would need to do more experiments.

> A pattern is correctly described. The student needs to describe the pattern as a relationship, such as the higher the number of carbon atoms the more energy is transferred, however 'benefit of doubt' would be given here. The answer is not complete though as the further data to justify the relationship needs to ensure that all other variables have been kept the same and that the mass of fuel burned is taken into account.

End of chapter questions

Getting started

1 What unit is temperature measured in? Circle your answer.

    1 Mark

    g   J   °C

2 Which piece of apparatus is used to measure temperature?

    1 Mark

    a thermometer   **b** calorimeter   **c** pipette

3 Describe **two** uses that exothermic or endothermic reactions can be applied to.

    2 Marks

4 Two solutions are added together. The temperature goes down. What type of reaction is this?

    1 Mark

    a addition   **b** endothermic   **c** thermal decomposition

5 100 g water was heated by 1 g fuel. Then 50 g water was heated by 1 g of the same fuel. Which water sample had the biggest temperature change?

    1 Mark

6 Sam and Alex use a fuel to heat water and measure the temperature change of the water. Explain why they put four hardboard mats around their fuel container and the water container.

    2 Marks

7 Match the result readings to the experiment.

    2 Marks

°C	20	28	34	36	37
cm³	0	15	20	24	26
°C	21	18	16	15	14

rate of reaction by gas volume
endothermic reaction
exothermic reaction

Going further

8 What unit is energy measured in?

    1 Mark

9 In an endothermic reaction which best describes the energy level of the reactants?

    a it is lower than the energy level of the products

    b it is higher than the energy level of the products

    c it is at the same energy level as the products

    1 Mark

10 Explain where you would mark the 'activation energy' of a reaction on an energy profile.

    2 Marks

11 Describe **one** experiment you would do to show that 2 g fuel A gives out more energy when burned than 2 g of fuel B.

    4 Marks

12 When crisps are burned to heat water (at 20 °C), the final temperatures are different. Suggest two questions you should ask to check if these results fairly compare the energy from the crisps.

    2 Marks

Weetos	Crispos	Liteos
24 °C	26 °C	22 °C

More challenging

13 What has to be overcome for a reaction to take place?

1 Mark

14 When adding two reactants in a beaker to measure temperature changes, explain why it is important to minimise heat loss.

1 Mark

15 When citric acid is added to sodium hydrogencarbonate solution the temperature decreases. Explain why, in terms of breaking and making bonds.

2 Marks

16 Use the collision theory to explain what happens when a reaction begins.

2 Marks

Most demanding

17 Using the data in question 19, calculate the bond energies for these compounds:

a carbon dioxide $O=C=O$

b water $H-O-H$

c ethane H_3C-CH_3

4 Marks

18 Explain why a reaction can be exothermic when bond breaking is an endothermic process.

2 Marks

19 These bond energies are given in kJ/mol.

C–H	C=O	O=O	H–O	C–C
412	805	498	465	368

Calculate the overall energy change during the burning of methane:

4 Marks

$$CH_4 + 2O_2 \rightarrow CO_2 + 2H_2O$$

20 Use the data from the table in question 19 to calculate whether propane C_3H_8 would give out more or less energy than methane when burned.

4 Marks

Total: 40 Marks

THE RATE AND EXTENT OF CHEMICAL CHANGE

IDEAS YOU HAVE MET BEFORE:

MEASURING REACTIONS

- Mass is measured in grams using a digital balance.
- Some chemical reactions show a colour change.
- If a gas is given off in a reaction, its volume can be measured.

CHEMICALS USED UP IN REACTIONS

- If more of a chemical is used, more of another is needed.
- If you use too much solid you have to filter it off.
- The word for too much is 'excess'.

RATES OF REACTION

- Some reactions are very slow, such as rusting.
- Some chemical reactions are very fast.
- Some chemical reactions are so fast they produce explosions.

CONTROLLING RATES OF REACTIONS

- Small pieces of sugar dissolve quicker than lumps.
- Boiling water will cook eggs quicker.
- Concentrated acid is more hazardous than dilute acid.

REVERSIBLE REACTIONS AND EQUILIBRIUM

- Copper sulfate can change from blue to white.
- Ammonia is a gas.
- When making chemicals we want to make as much as possible.

IN THIS CHAPTER YOU WILL FIND OUT ABOUT:

HOW CAN WE MEASURE REACTION RATES?

gas syringe

- Amount reacted can be measured by collecting gas volumes.
- Amount reacted can be measured by loss of mass.
- Amount reacted can be measured by a colour change.

WHAT AFFECTS THE END OF A REACTION?

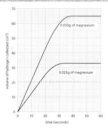

- Reactions are usually faster at the start, and slow down.
- The reaction stops when one reactant is used up.
- The reactant that is used first is the 'limiting reactant'.

HOW CAN WE CALCULATE RATES OF REACTION?

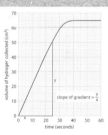

- Mean reaction rate can be calculated as product divided by time.
- Rates of reaction can be calculated from tangents on graphs.
- Tangents can be drawn at any point on the graph.

WHAT FACTORS AFFECT RATES OF REACTIONS?

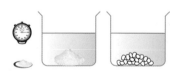

- Surface area, pressure and concentration affect reaction rate.
- Increasing temperature causes more successful collisions.
- Catalysts lower activation energy, making reactions faster.

HOW CAN REACTIONS BE IN EQUILIBRIUM?

- A reversible reaction has a forward reaction and a backward reaction.
- Increasing pressure can change the position of equilibrium.
- Le Chatelier's principle can be applied to systems in equilibrium.

Measuring rates

Learning objectives:

- explain how to measure the amount of gas given off in a reaction
- explain how to measure the rate of a reaction
- read data from graphs to interpret stages of a reaction.

KEY WORDS

balance
gas syringe
measuring cylinder
volume

It is difficult to measure the rate of very fast or very slow reactions. However, the rate of some reactions can be measured easily. Here are some of the experimental methods to measure reaction rates.

Ways to measure reaction times

There are many ways to measure the time taken for a chemical reaction. Three of these ways are by measuring:

- the loss in mass of reactants over time
- the volume of gas produced over time
- the time for a solution to become opaque or coloured.

During the reaction between calcium carbonate and dilute acid a gas is given off.

calcium carbonate + hydrochloric acid

$\rightarrow$ calcium chloride + water + carbon dioxide

If this is done in a flask, the carbon dioxide gas will escape. The mass of the flask's contents will go down. The decrease in mass can be measured on a digital **balance**.

The decrease in mass can be measured each minute and plotted on a graph.

The time for the reaction between magnesium and dilute hydrochloric acid can also be measured by this method.

magnesium + hydrochloric acid $\rightarrow$ magnesium chloride + hydrogen

Look at Figure 6.1 to answer questions 1 and 2.

1 **Estimate the decrease in mass after two minutes.**

2 **During which minutes was the graph steepest? Suggest why this is.**

Volumes of gas

It is often easier to measure the volume of a gas made rather than very small changes in mass.

Figure 6.2 shows three ways of collecting gas given off in a reaction using an upturned **measuring cylinder**, an upturned burette or using a '**gas syringe**'.

The **volume** of hydrogen collected in the gas syringe can be measured every 10 seconds.

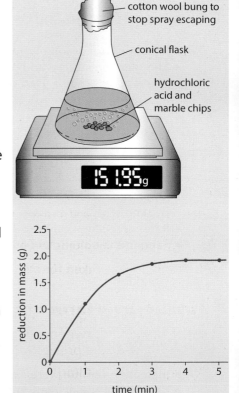

cotton wool bung to stop spray escaping

conical flask

hydrochloric acid and marble chips

Figure 6.1 Measuring loss in mass every minute

KEY SKILLS

Remember to make sure you have a cotton wool plug in the flask to prevent damage to the balance from the fizzing contents.

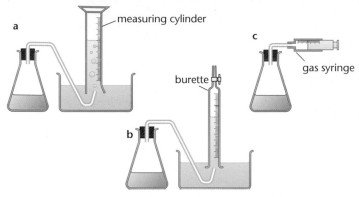

measuring cylinder

a

c

gas syringe

burette

b

Figure 6.2 Three ways of collecting and measuring volumes of a gas

The results of the experiment can be plotted on a graph, as shown in Figure 6.3. The slope of the line shows how fast the reaction is. The steeper the slope is, the faster the reaction.

Use Figure 6.3 to answer questions 3 and 4.

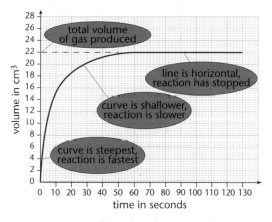

Figure 6.3 Gas collected and measured every 10 seconds

3 Determine the volume of gas produced in the first 20 seconds.

4 State the time taken for the reaction to stop.

Measuring degree of opaqueness

In the reaction between sodium thiosulfate and hydrochloric acid a precipitate of sulfur is slowly produced.

$$Na_2S_2O_3(aq) + 2HCl(aq) \rightarrow 2NaCl(aq) + H_2O(l) + SO_2(g) + S(s)$$

The flask is put over a paper with a large X marked on it. When the reactants are mixed the time taken for the cross to 'disappear' is measured with a timer.

5 The concentration of sodium thiosulfate can be varied. Explain how you could adapt the experiment to measure the effect of this on rate.

6 Give **two** reasons why the method may lack accuracy and lead to errors. Suggest **one** modification to improve the accuracy for each of these.

DID YOU KNOW?

If an upturned burette or measuring cylinder is used it must be filled with water before it is turned upside down. The volume is read off the scale on the side.

Figure 6.4 The precipitate of sulfur forms slowly to obscure the X. The time is taken when the cross just 'disappears'.

KEY CONCEPT

Limiting reactants and molar masses

Learning objectives:

- identify which reactant is in excess
- explain the effect of a limiting quantity of a reactant on the amount of products
- calculate amount of products in moles or masses in grams.

KEY WORDS

limiting
excess
mole
directly
  proportional

When you watched your teacher put some sodium on water, they only used a tiny amount with a great deal of water. This was for safety, but everyone always wants them to use more. They only use a small amount to *limit* the reaction. When it's gone, it's gone.

HIGHER TIER ONLY

Reactants in excess

In a reaction between magnesium and dilute hydrochloric acid a small amount of one of the reactants is used and a lot more of the other reactant is used. We need to decide:

- when to use a lot more of one reactant
- what is 'a lot more'.

For example, when making the salt, magnesium chloride, all the acid needs to be used up, as it is hard to separate the magnesium chloride as a solid from the acidic solution.

$$Mg(s) + 2HCl(aq) \rightarrow MgCl_2(aq) + H_2(g)$$

It is much easier to have too much magnesium metal, use up all the acid reacting with the metal, then filter off the solution from the **excess** magnesium solid.

In this case the **limiting** reagent is the hydrochloric acid and the excess is magnesium.

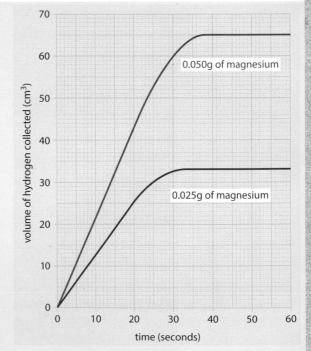

Figure 6.5 How much more hydrogen is formed if the amount of magnesium is doubled?

On the other hand if we are investigating the rate of reaction between magnesium and hydrochloric acid, we will want to use up all of the magnesium, as it is easier to measure the mass of the solid than adjust the concentration and volume of acid. So we use excess acid and a known amount of magnesium.

Magnesium is used up by the end of the reaction so it is known as the *limiting reactant*.

Look at the graph in Figure 6.5.

In this reaction, both Mg and HCl are the reactants.
To find the rate of reaction we limit the amount of magnesium.

1 Use construction lines to find the rate of reaction when 0.050 g of magnesium is used (blue line).

2 Determine the volume of hydrogen formed when:

  a 0.025 g of magnesium is used.

  b 0.050 g of magnesium is used.

  c Describe the relationship between the amount of reactant used and the amount of product formed.

3 Predict the volume of hydrogen if 0.0125 g of magnesium is used. Sketch this reaction line on a copy of the graph.

> **REMEMBER!**
>
> The limiting reactant *limits* the amount of product that can be made.

How much product?

The amount of product is **directly proportional** to the amount of limiting reactant used. So, if the amount of this reactant doubles, the amount of product doubles. The more reactant that is used, the more product it forms.

4 Explain what is meant by 'limiting reactant'.

5 Predict what will happen to the amount of product if the limiting reactant is quadrupled.

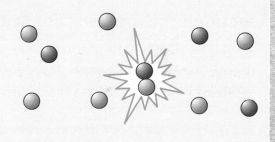

Figure 6.6 Which colour is the limiting reactant in this model of colliding particles?

Molar masses and moles again

$$Mg(s) + 2HCl(aq) \rightarrow MgCl_2(aq) + H_2(g)$$

Magnesium is the limiting reactant.

- If 24 g of magnesium had been used in the reaction we should have made 95 g of $MgCl_2$
- If 12 g of magnesium had been used in the reaction we should have made 47.5 g of $MgCl_2$
- If 6 g of magnesium had been used in the reaction we should have made 23.75 g of $MgCl_2$

6 Work out the mass of magnesium chloride that could be made from 3 g of magnesium.

7 Explain why the numbers in the first row above (24 g Mg and 95 g of $MgCl_2$) were chosen.

8 In the reaction:

  $$Mg(s) + 2HCl(aq) \rightarrow MgCl_2(aq) + H_2(g)$$

  If 2 moles of $MgCl_2$ are needed,

  a determine how many moles of Mg must be used at the start

  b determine the minimum number of moles of HCl that must be used for the reaction.

> **DID YOU KNOW?**
>
> If the mass of the atomic mass of an element is measured out in grams this will be the same as the mass of one **mole** (the molar mass).

Calculating rates

KEY WORDS

axes
construction lines
gradient
rate of reaction

Learning objectives:

- calculate the mean rate of a reaction
- draw and interpret graphs of reaction times
- draw tangents to the curves as a measure of the rate of reaction.

We can measure the progress of a reaction by collecting gas, measuring the loss of mass or the degree of opaqueness. If we do this in set time intervals we can then calculate the rate of a reaction.

Reaction time and reaction rate

In a chemical reaction reactants are made into products.

reactants → products

The **rate of reaction** measures how much product is made each second.

Some reactions are very fast and others are very slow:

- Rusting is a very slow reaction.
- Burning and explosions are very fast reactions.

The *reaction time* is the time taken for the reaction to finish.

The shorter the reaction time is, the faster the reaction.

The *rate of reaction* shows how much product is formed in a fixed time (or how fast a reactant disappears).

So either: mean rate of reaction = $\dfrac{\text{quantity of reactant used}}{\text{time taken}}$

or mean rate of reaction = $\dfrac{\text{quantity of product formed}}{\text{time taken}}$

Figure 6.7 Slow and fast reactions

DID YOU KNOW?

The rate of a reaction is an important factor in running industrial processes.

1. Describe the following reactions in terms of rate and time taken.

   a Glue setting

   b Methane combusting

   c Diamond and oxygen at room temperature

2. Hydrochloric acid was added to lumps of calcium carbonate. In the first 50 seconds, 30 cm³ of gas was collected. Calculate the mean rate of reaction.

Changes of rate

The rate changes during a reaction.

Reactions are usually faster at the start, and then they slow down as the reactants are used up.

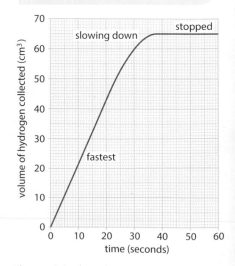

Figure 6.8 Changing rates

An example is the reaction between magnesium ribbon and hydrochloric acid.

$$Mg + 2HCl \rightarrow MgCl_2 + H_2$$

Hydrogen is a gas so it is easy to measure how much is produced.

Look at Figure 6.8. The steeper the gradient (slope) on the graph is, the faster the reaction. As the **gradient** decreases, the reaction slows down. When the line is horizontal the reaction has finished, so no more product is being made. This is because one of the reactants has been used up.

The rate of reaction can be worked out from the gradient of a graph.

The gradient is found by drawing **construction lines**. Choose a part of the graph where there is a straight line (not a curve). Measure the values of y and x and remember to use the scale of the graph carefully. Then divide y by x. Determine the units from those given on the **axes**.

In Figure 6.9 the gradient $= y/x = 50/25 = 2\ cm^3/s$

It is very important to check the units you are using.

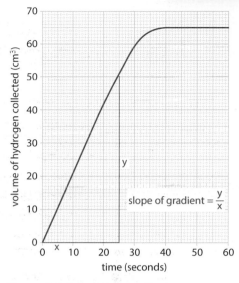

Figure 6.9 Calculating gradients

Quantity of reactant or product	Unit of quantity	Unit of rate
Mass	g	g/s
Volume	cm^3	cm^3/s
Moles	mol	mol/s

3 30 cm^3 of hydrogen is made in 20 seconds. Calculate the reaction rate.

4 Look at Figure 6.10. Draw construction lines to calculate the gradient of the red line.

5 Explain the difference in gradients between the red and blue lines in Figure 6.10.

> **REMEMBER!**
>
> Sometimes it is easier to measure the mass of a product formed. The rate of reaction is then measured in g/s or g/min.

HIGHER TIER ONLY

Rates at specific times

The rate changes during a reaction. The tangent to the curve can be chosen at a specific point on a graph and the gradient calculated as a measure of the rate of reaction at a specific time.

6 Use Figure 6.10 to calculate the gradients to determine the rates of reaction at (a) 20 seconds and (b) 40 seconds on both the red and blue lines.

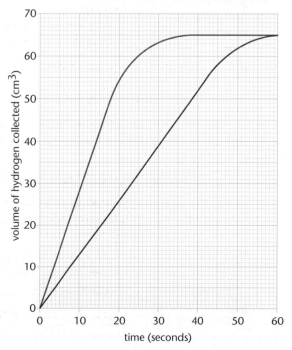

Figure 6.10 Different gradients

Factors affecting rates

Learning objectives:

- identify which factors affect the rate of reactions
- explain how changes of surface area affect rates
- explain how rates are affected by different factors.

KEY WORDS

catalyst
concentration
pressure
temperature

There are many ways to alter the rate of a reaction. Gases reacting together can be put under pressure to put the particles nearer to each other. Catalysts can be used to help particles have a surface on which to bind. Some of these ways are easy to investigate.

Factors affecting the rate of reaction

Factors that affect the rates of chemical reactions include the:

- **temperature**
- **concentrations** of reactants in solution
- **pressure** of reacting gases
- surface area of solid reactants
- presence of **catalysts**.

In the reaction between sodium thiosulfate and hydrochloric acid a precipitate of sulfur is made. This happens slowly enough to measure the time that it takes to make a large X under the reaction flask 'disappear' (see Figure 6.4 in topic 6.1).

Figure 6.12 shows the results when the reaction was carried out at *different temperatures*.

1 Determine the time taken for the reaction to reach the end point at 35 °C.

2 What is the trend as the temperature increases?

Changing concentrations

The sodium thiosulfate and hydrochloric acid reaction can be used to investigate the effect of changing concentrations. The temperature would be kept the same each time, but different *concentrations* of sodium thiosulfate would be used.

3 The concentration of sodium thiosulfate was increased without changing the temperature. Describe what would happen to the time taken for the cross to disappear and the reaction rate.

4 The effect of concentration on rate of reaction between calcium and hydrochloric acid was investigated.

	Temperature / °C	Concentration of HCl / mol/dm³
Experiment 1	20	0.1
Experiment 2	25	0.2

Suggest whether the investigation was valid.

KEY SKILLS

Remember that if you change the temperature (one factor) you must keep everything else the same.

Figure 6.11 The time is taken when the cross just disappears. This is repeated at different temperatures.

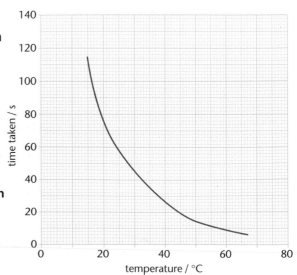

Figure 6.12 Temperature against time taken for the cross to disappear

Surface area to volume ratio

The calcium carbonate and dilute hydrochloric acid reaction can be used to investigate the effect of changing the surface area of solid reactants. The acid concentration would be kept the same each time but the size of the calcium carbonate pieces would change, while keeping the mass the same.

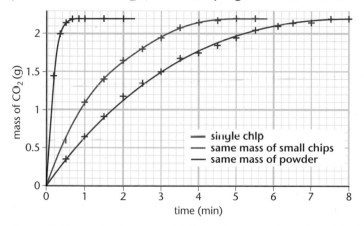

Figure 6.13 Graph of mass of CO_2 lost using three sizes of marble chips

The reason that the rate of reaction changes with the size of pieces of a reacting solid is because there is a bigger surface area that the volume of acid can attack, if the chip is cut into smaller pieces. Look at Figure 6.14. The large cube has six large faces for attack. If the cube is divided, those six faces remain but there are other surfaces that were 'locked inside' that are now available for attack so the reaction of the acid can produce more CO_2 per minute.

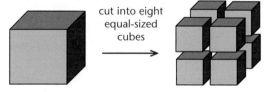

2 cm × 2 cm × 2cm cube   1 cm × 1 cm × 1 cm each cube

Figure 6.14 Increasing surface area

5 Use Figure 6.13 to determine the mass of CO_2 lost using a single chip after three minutes.

6 Describe the pattern linking the size of the chip and the time taken to produce carbon dioxide.

7 Work out the mean rate of reaction, including units, for each of the three reactions above.

8 Explain why the same mass of CO_2 was produced in each of the three experiments.

9 Describe three ways that the rate of the Mg/HCl reaction could be altered. Would these increase or decrease the rate?

KEY INFORMATION

Remember that if you change the concentration of sodium thiosulfate you must keep the volume the same. You must also keep the concentration and volume of hydrochloric acid the same and the temperature the same too.

DID YOU KNOW?

You can either measure the volume of carbon dioxide given off using a gas syringe or the loss in mass using a digital balance (see topic 6.1).

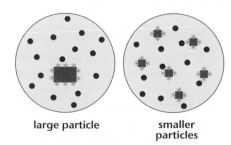

large particle   smaller particles

Figure 6.15 Increased surface area can be attacked by acid.

REQUIRED PRACTICAL

Investigate how changes in concentration affect the rates of reactions by a method involving the production of a gas and a method involving a colour change

KEY WORDS

concentration
volume
rate
tangent

Learning objectives:

- use scientific theories and explanations to develop a hypothesis
- plan experiments to test the hypothesis and check data
- make and record measurements using gas syringes
- evaluate methods and suggest improvements and further investigations.

Rates of reaction are very important for industrial processes and many chemical engineers are involved in controlling rates on large scale processes. Several factors affect the rate of a reaction. Are you able to spot the trends if one variable is changed?

Developing a hypothesis

A number of different skills are needed to carry out a *full* investigation. First a hypothesis needs to be developed from previous knowledge or observations. Consider an investigation that shows the **rate** of reaction of acid with magnesium increasing when the **concentration** of the acid increases. Your idea, from this conclusion, may be 'the more particles of acid there are, the more chances of successful particle collisions taking place'.

After this evaluation another question could be: what other variable change might increase the number of successful particle collisions? Could increasing temperature do this?

1. **Suggest a hypothesis.**

2. **The reaction you would follow is acid on carbonate. Identify the variable that should be changed.**

3. **Suggest which variables need to be kept the same.**

Planning the investigation: Figure 6.16 shows a diagram of the apparatus that might be used to follow the reaction of acid on magnesium.

Think about these questions:

4. **Explain why a gas syringe is used.**

5. **What would need to happen quickly when the magnesium is put into the flask?**

> ❗ These pages are designed to help you think about aspects of the investigation rather than to guide you through it step by step.

DID YOU KNOW?

You can make an experiment so that a colour change happens at an exact time, by adjusting the concentration of the solutions.

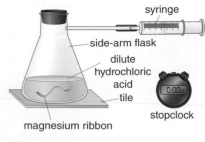

Figure 6.16

Making and recording measurements:

Measurements need to be accurately made, and the data collected, recorded and processed. Graphs of the results need to be drawn and, if possible, **tangents** to the curves drawn to calculate rates at different points of the reaction.

Analysing the results

Sam and Alex had investigated the effect of temperature on the reaction between acid and calcium carbonate (using the same apparatus as above) and had gathered these results:

Time in secs	Volume of gas collected from acid at specific temperature in cm³						Rate of reaction in cm³/sec at 20°C
	20°C	20°C	30°C	30°C	40°C	40°C	20°C
0	0	0	0	0	0	0	
10	12	11	24	45	49	50	
20	23	21	45	47	64	64	
30	34	35	62	61	64	64	
40	52	51	64	64	64	64	
50	63	62	64	64	64	64	
60	64	64	64	64	64	64	

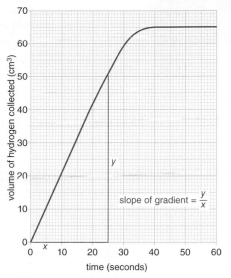

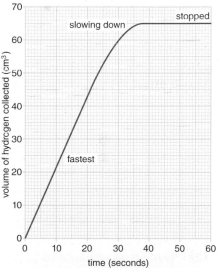

6 Identify the relationship that Sam and Alex were exploring.

7 Identify the anomaly in the data. Suggest what Sam and Alex should do with this data point.

8 Plot three lines on a graph of volume of gas collected against time, one line for each temperature.

9 Suggest what their conclusion might be.

10 Work out the average rate of reaction over 40 seconds for the 20 °C reaction.

Evaluating the experiment

Sam and Alex were testing the hypothesis: 'does an increase in temperature increase the rate of reaction between acid and calcium carbonate?' Using their conclusion in question 9 reflect on how well the data supports their conclusion.

11 Explain whether the evidence supports their conclusion or whether someone could use the same data to support a different conclusion.

12 Explain how they made sure that their data was repeatable.

13 Explain how the apparatus for their investigation and their method of use ensured that their findings were *valid*.

14 Suggest **one** way the apparatus or method could have been improved.

15 Suggest **one** way they could find out if their specific hypothesis could be extended more generally.

16 Explain, using the particle model, why their conclusion may be applied to this reaction and other reactions.

KEY SKILL

It is important to be very methodical in recording results. Tables need to be clearly labelled.

Factors increasing the rate

Learning objectives:

- analyse experimental data on rates of reaction
- predict the effects of changing conditions on rates of reactions
- use ideas about proportionality to explain the effect of a factor.

KEY WORDS

excess
analyse
graph
steeper

We have already seen how to measure the change of the rate of reactions and which factors can be altered to increase the rate. We have also developed a theory to explain these changes. We now need to ensure that we present the results so that evidence can be analysed.

Analysing data

We have seen how we can collect the results of a reaction to measure its rate. Look at the results of collecting a gas in a gas syringe.

We can see that more gas is given off in 0 to 20 seconds than in 20 to 40 seconds. The rate of gas given off has become less. The rate has decreased.

Look at Figure 6.17 to answer the questions.

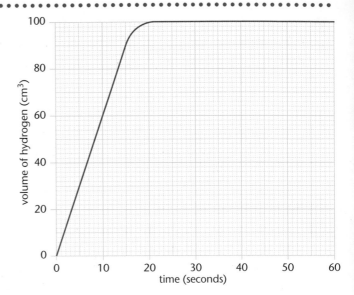

Figure 6.17 Collecting gas between 0 and 60 seconds

1. When does the reaction stop?
2. Identify **two** substances that produce hydrogen when they react.
3. Explain what happens to the rate of gas given off each second between 21 and 60 seconds.
4. Calculate the average rate during the first 10 seconds.

Increasing the rate of reaction

We have seen that we can make the reaction faster by:

- increasing the concentration
- increasing the temperature
- increasing the pressure (of gases)
- increasing the surface area (of a solid).

From the **graphs** in Figures 6.18 and 6.19, we see how the rate of reaction changes with increasing (a) the concentration (b) the temperature.

We can see that there was 30 cm³ of gas made in the first 10 seconds when the high concentration acid is used and increasing the concentration gives a **steeper** curve.

DID YOU KNOW?

In Figure 6.18 the red line shows the results of the experiment with dilute hydrochloric acid and the blue line shows results with acid that is more concentrated.

The gradient of the blue line is greater than that of the red line. This shows that the rate of reaction is faster when more concentrated acid is used.

In Figure 6.19 the blue line shows the results with the acid solution at a temperature of 30 °C. Its gradient is greater than that for the red line, which describes the experiment in which the acid solution was at a temperature of 20 °C.

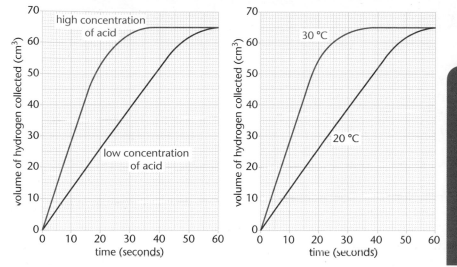

Figure 6.18 Increasing the concentration   Figure 6.19 Increasing the temperature

KEY INFORMATION

Figures 6.18 and 6.19 show the results from four different experiments. The total volume of hydrogen produced during all the experiments is the same. This is because **excess** acid and the same mass of magnesium are used.

5) Fill in Tables 6.1 and 6.2 and explain what the data shows.

6) Determine the total volume of gas made. Explain why it is the same for the reactions at both 20 °C and 30 °C.

7) Predict and sketch what the curved line will look like if the reaction is carried out at 40 °C.

Changing the mass

We can alter the rate of reaction by changing the *factors* but the amount of product will remain the same.

However, if we reduce the amount of one reactant used, then less product will be formed.

8) Look at Figure 6.20. Explain what the data shows about:

   a the amount of product formed in both reactions

   b the rate of reaction in both reactions.

9) Predict and sketch what the curved line will look like if the reaction is carried out with 0.0495 g of magnesium.

10) An investigation was carried out into the rate of reaction between solid calcium carbonate, $CaCO_3$, and nitric acid. The mass of calcium carbonate was kept the same, as was the temperature. The results are shown in the table below.

	Time taken to produce 10 cm³ of CO_2 / seconds	Total volume of CO_2 produced / cm³
Experiment 1	27	40
Experiment 2	14	80

   a Write a balanced equation for the reaction. Include state symbols.

   b Explain the results of experiments 1 and 2.

Volume of gas made after 10 seconds	
Low acid concentration cm³	High acid concentration cm³
	30

Table 6.1

Volume of gas made after 10 seconds	
Low temperature °C	High temperature °C
	28

Table 6.2

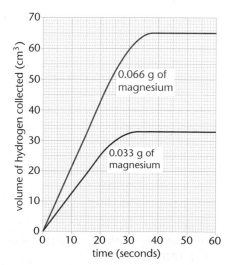

Figure 6.20 Changing the mass of reactant

Collision theory

Learning objectives:

- describe a reaction in terms of particles colliding
- explain the effects of changes of factors on rates of reaction using collision theory
- describe activation energy.

KEY WORDS

activation energy
collision
collision
   frequency

We know that we can change the rate of reaction by altering the concentration, temperature, pressure and other factors. We know how we can measure the changes in reactions. Now we need to be able to *explain* what is happening by developing a theory – collision theory.

Collisions

Chemical reactions take place when reactant particles collide with each other and form new products.

Not all collisions are successful, but a reaction takes place when a **collision** is successful.

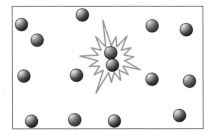

Figure 6.21 Collisions and successful collisions between reacting particles

To make the reaction slower, the number of successful collisions needs to decrease.	To make the reaction faster, the number of successful collisions needs to increase.

A reaction can be made to go faster by:

- increasing the temperature
- increasing the concentration (of reactants in solution)
- increasing the pressure (if the reactants are gases)
- increasing the surface area of solid reactants
- the presence of catalysts.

All of these changes allow more successful collisions. The more successful collisions there are, the faster the reaction.

① **Explain why increasing the pressure on two gases reacting will make the reaction go faster.**

Activation energy

Collision theory explains that chemical reactions can occur only when reacting particles hit or collide with each other with sufficient energy.

It explains how various factors affect rates of reactions in terms of particles colliding.

As the concentration increases the particles become more crowded.

For example, at the low concentration in Figure 6.23 there are four particles of A. At the higher concentration there are ten particles of A in the same volume.

KEY INFORMATION

Remember that increasing the surface area of solid marble chips by breaking them down into powder will allow more surface area for the particles of acid to collide with and react.

small surface area

large surface area

Figure 6.22 Increasing surface area increases the chances of successful collisions.

low concentration

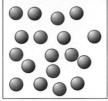
high concentration

⬤ reacting particle of substance **A**
⬤ reacting particle of substance **B**

Figure 6.23 Increasing concentration increases chances of successful collisions.

The particles are more crowded in the same volume, there are more collisions, and so there are more chances of successful collisions and the rate of reaction increases.

Increasing the temperature also increases the rate of reaction.

Particles move faster as the temperature increases. The reacting particles have more kinetic energy and so the number of successful collisions increases and the rate of reaction increases.

For a successful collision to occur each particle must have sufficient energy to react. This minimum amount of energy that particles must have to react is the **activation energy**.

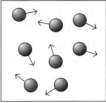
low temperature

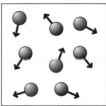
high temperature

⬤ reacting particle of substance **A**
⬤ reacting particle of substance **B**

Figure 6.24 Increasing temperature increases chances of successful collisions.

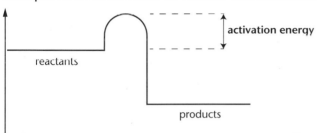

activation energy

reactants

products

Figure 6.25 There is a minimum energy that has to be overcome for colliding particles to react. This is the activation energy.

2 **Explain, using collision theory, how increasing the temperature of the reaction between magnesium ribbon and dilute hydrochloric acid increases the rate of reaction.**

Collision frequency

It is not just the number of collisions that determines the rate of a reaction, it is the **collision frequency**.

Collision frequency describes the number of successful collisions between reactant particles that happen each second. The more successful collisions there are per second, the faster the reaction.

3 **Explain how increasing the concentration of acid increases the rate of reaction of:**

$$CaCO_3(s) + 2HCl(aq) \rightarrow CaCl_2(aq) + H_2O(l) + CO_2(g)$$

4 **A typical collision frequency for gas particles is 5×10^{10} per second. Calculate the time taken for 1×10^6 collisions to occur.**

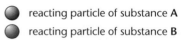

DID YOU KNOW?

There are ways to reduce the activation energy using catalysts.

Catalysts

Learning objectives:

- identify catalysts in reactions
- explain catalytic action
- explain activation energy.

KEY WORDS

activation energy
catalytic
catalyst

Scientists are always searching for new catalysts that can make a specific reaction faster, more energy efficient and more cost effective. Catalysts are now always used in catalytic converters in car exhausts to change polluting exhaust gases into less harmful gases.

Catalysts

Catalysts are substances that change the speed of a chemical reaction but are not used up during that reaction.

Different reactions need different catalysts.

A catalyst:

- is a substance added to a chemical reaction to make the reaction go faster (or to start)
- is only needed in small amounts
- remains unchanged at the end of a reaction.

When a catalyst is added to a reaction the *same amount* of product is produced but in a much shorter time.

For example:

Sam and Alex investigate the reaction between zinc and sulfuric acid. Zinc sulfate solution and hydrogen gas are formed.

They want to find a catalyst for this reaction.

They measure the time it takes to collect 100 cm³ of hydrogen in a gas syringe.

They add another substance each time.

Their results are shown in Table 6.3.

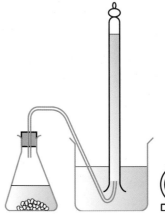

Figure 6.26 Measuring the time taken to collect 100 cm³ of hydrogen

KEY INFORMATION

Remember that when zinc is added to a solution of copper chloride the reaction displaces *copper* from the solution.

Experiment number	Substance	Time to collect 100 cm³ of gas (seconds)	Colour of substance at the start	Colour of substance at the end
1	no substance	150	—	—
2	magnesium chloride	150	white	white
3	copper chloride	15	green	pink
4	copper powder	25	pink	pink

Table 6.3

1 Explain whether each substance is a catalyst or not.

Specific catalysts

Catalysts change the rate of chemical reactions but are not used up during the reaction.

Some other properties of a catalyst are that:

- only a small mass of catalyst is needed to catalyse a large mass of reactants
- they are specific to one single reaction.

Most catalysts only make a specific reaction faster. They do not make all reactions faster.

Examples of catalysts that are widely used are:

- vanadium pentoxide in the Contact process
- iron in the Haber process
- zeolites in the cracking of long-chain hydrocarbons
- rhodium-based catalysts in a **catalytic** converter.

2 **Write a balanced symbol equation for the reaction between zinc and sulfuric acid with a copper catalyst.**

Activation energy pathways

A catalyst does not increase the number of collisions per second. Instead, it works by making the collisions that take place more *successful*.

For reactions to take place the successful collisions have to have enough energy to overcome the **activation energy**.

Catalysts create a new pathway for the reaction. This new pathway has a lower *activation energy*. Catalysts therefore increase the rate of reaction by providing a different activation energy pathway.

A reaction profile for a catalysed reaction can be drawn as in Figure 6.27.

Enzymes are very important in maintaining the life and health of organisms and need special conditions to work effectively.

3 **Use a reaction profile diagram to explain how copper sulfate allows hydrogen peroxide particles to break down more quickly.**

4 **Hydrogen peroxide is a strong oxidising agent and is used as bleach and disinfectant. When it decomposes, it produces oxygen. The activation energy for the decomposition has been measured and the results are shown in the table.**

a **Write a balanced equation for the decomposition of hydrogen peroxide.**

b **Identify the most effective catalyst and explain how it works.**

c **Suggest a value for the activation energy without a catalyst. Justify your answer.**

d **Hydrogen peroxide is produced in cells in the body. Explain why catalase is also present in cells.**

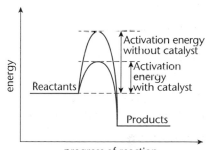

Figure 6.27 Catalysts lower the activation energy of a reaction.

DID YOU KNOW?

In a biological system it is an *enzyme* that acts as a catalyst. Enzymes also work by lowering the activation energy pathway.

Catalyst	Activation energy in kJ/mol
None	?
Catalase enzyme	23
Platinum, Pt	49
Iodide ions, I^-	56
Manganese(IV) oxide, MnO_2	58

Reversible reactions and energy changes

Learning objectives:

- identify a reversible reaction
- explain how energy changes occur in reversible reactions
- consider changing the conditions of a reversible reaction.

KEY WORDS

backwards
exothermic
endothermic
forwards
reversible

In precipitation reactions, it seems that when the two solutions are added a reaction happens immediately and a new product is seen – the precipitate is made. This does not happen in all reactions. The reaction may need certain conditions to happen and the reaction may go backwards as well.

Reversible reactions

In a chemical reaction, reactants react to make products.

$$reactants \rightarrow products$$

However, in some chemical reactions, the products of the reaction can react to make the original reactants again.

These reactions are called **reversible** reactions and are represented with a double half-headed arrow.

$$reactants \rightleftharpoons products$$

There is a **forward** reaction and a **backward** reaction. These take place at the same time.

An example of a reversible reaction is the heating and cooling of ammonium chloride.

$$ammonium\ chloride \underset{cool}{\overset{heat}{\rightleftharpoons}} ammonia + hydrogen\ chloride$$

The white ammonium chloride decomposes to two colourless gases. When the gases are cooled they re-form into white ammonium chloride.

Figure 6.28 Heating ammonium chloride is a reversible reaction.

1. **Suggest whether the following are reversible or irreversible.**

   a Wood burning

   b Water melting

   c Formation of ammonia from hydrogen and nitrogen

2. **Explain whether dissolving sodium chloride in water is reversible or not.**

Energy changes in reversible reactions

Another example of a reversible reaction is the heating of blue copper sulfate.

Figure 6.29 Blue copper sulfate turns white on heating and returns to blue if water is added.

At first it is a hydrated blue crystal. When heated it loses the water of crystallisation to become white anhydrous copper sulfate. The reaction is represented by:

hydrated copper sulfate (blue) $\underset{\text{exothermic}}{\overset{\text{endothermic}}{\rightleftharpoons}}$ anhydrous copper sulfate (white) + water

The forward reaction takes place when the crystals are heated. Energy is transferred in so it is an **endothermic** reaction.

The backward reaction takes place when the water is added to the anhydrous copper sulfate. Energy is transferred out so it is an exothermic reaction.

The same amount of energy is transferred out as the amount of energy transferred in.

3 **Explain what the term 'endothermic' means.**

DID YOU KNOW?

Anhydrous means 'without water'.

HIGHER TIER ONLY

Changing direction

A reversible reaction has a forward reaction and a backward reaction that take place at the same time.

$$A + B \rightleftharpoons C + D$$

What do you think happens if the forward and backward reactions take place at the same rate? We will meet this idea in the next pages.

However, if a reversible reaction is **exothermic** in one direction, it is endothermic in the opposite direction. The same amount of energy is transferred in each case.

The direction of reversible reactions can be changed. This can be done by changing the conditions.

For example, nitrogen and hydrogen react to form ammonia.

nitrogen + hydrogen $\rightleftharpoons$ ammonia

$$N_2 + 3H_2 \rightleftharpoons 2NH_3$$

The ammonia made in the forward reaction breaks down to form nitrogen and hydrogen again.

If the pressure is increased the reaction will move to make more product. If the pressure is decreased there will be more backward reaction instead.

DID YOU KNOW?

Nitrogen and hydrogen are used to make ammonia in the Haber process. Ammonia is needed for another process to make fertilisers for growing better crops.

4 *Suggest* why increasing the pressure in the reversible reaction

$$N_2 + 3H_2 \rightleftharpoons 2NH_3$$

may favour the forward reaction. Think about particle theory.

5 An iron catalyst is used in the Haber process. Describe how it affects the energy change for the forward and reverse reactions.

Equilibrium

Learning objectives:

- describe how equilibrium is reached
- explain what happens to the forward and reverse reactions
- predict the effects of changes on systems at equilibrium.

In a reversible reaction, there is a forward reaction and a backward reaction. Sometimes these take place at the same time. The symbol $\rightleftharpoons$ is used to show that a reaction is reversible. When a balance is reached between the rates of these competing reactions, the reaction has reached equilibrium.

Reaching equilibrium

In a reversible reaction there is a forward reaction and a **reverse reaction**. What would be the outcome if these were happening at the same time?

Equilibrium is reached when the forward and reverse reactions occur at exactly the *same rate* but the apparatus must prevent the escape of the reactants and the products. This is called a closed system.

1 Explain why an open system with a gas escaping will not reach equilibrium.

2 Explain how a saturated solution of salt is at equilibrium in terms of rate.

Equilibrium position

At equilibrium:

- the rate of the forward reaction equals the rate of the backward reaction
- the concentrations of reactants and products do not *change*.

However, when the backward and forward rates balance and are equal, the *concentrations* of the reactants and products do not need to be equal. There is usually more of one than the other.

Reactants	$\rightleftharpoons$	Products
If the concentration of reactants is greater than the concentration of products, we say that the position of equilibrium is on the *left*.		If the concentration of the reactants is less than the concentration of the products, the position of equilibrium is on the *right*.

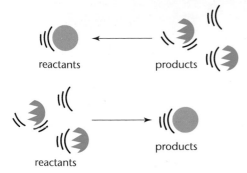

Figure 6.30 Position of equilibrium

Remember: this only works if the chemicals cannot escape.

Initially, the forward reaction rate is fast, but then it slows as the reactants are used up.

At the same time, the rate of the backward reaction increases as more products are available to react. Eventually the backward reaction has the same rate as the forward reaction. Equilibrium has been reached.

3 The reaction: $A + B \rightleftharpoons C + D$ is at equilibrium. The concentration of A + B is higher than the concentration of C + D. Suggest where the *position* of equilibrium lies.

HIGHER TIER ONLY

Le Chatelier's principle

The relative amounts of all the reactants and products at equilibrium depend on the conditions of the reaction.

If a system is at equilibrium and a change is made to any of the conditions, then the system responds to **counteract** the change. This is known as **Le Chatelier's principle**.

The effects of changing conditions on a system at equilibrium can be predicted using Le Chatelier's principle.

The reaction: $A + B \rightleftharpoons C + D$ is at equilibrium.

If the concentration of A + B is increased then the system will respond by making more C + D and so reducing the amount of A + B. The system will respond to the change by counteracting it.

4 Explain what will happen to the reaction

$$A + B \rightleftharpoons C + D$$

if it is in equilibrium and then the amount of C and D is increased.

5 Hydrochloric acid was added to bismuth(III) oxychloride and a clear solution formed. Describe and explain what you would see if water was added followed by hydrochloric acid.

$$BiOCl(s) + 2HCl(aq) \rightleftharpoons BiCl_3(aq) + H_2O(l)$$

DID YOU KNOW?

An example of a reversible reaction is making sulfur trioxide from sulfur dioxide and oxygen. Changing conditions can alter the amount of product made.

KEY INFORMATION

You will find out about how conditions change the position of equilibrium in the following topics.

Changing concentration and equilibrium

Learning objectives:

- identify reactants and products in a reversible reaction
- explain how changing concentrations changes the position of equilibrium
- interpret data to predict the effect of a change in concentration.

KEY WORDS

equilibrium
  position
concentration
dinitrogen
  tetroxide
reversible reaction

Reversible reactions are used in making the propellant for space rockets and for making fertilisers. These reactions can make more products by changing the position of equilibrium. By removing the product as it forms, more can be made. This means that space rockets can be powered and more food can be grown.

HIGHER TIER ONLY

Removing products

Some rocket propellants use **dinitrogen tetroxide** N_2O_4. Dinitrogen tetroxide is made from nitrogen dioxide (NO_2) in a **reversible reaction**.

$$2NO_2(g) \rightleftharpoons N_2O_4(g)$$

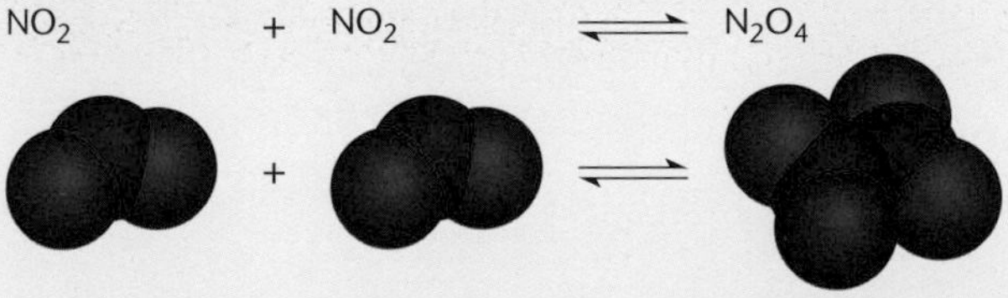

NO_2 + NO_2 $\rightleftharpoons$ N_2O_4

Figure 6.31 Dinitrogen tetroxide is made from nitrogen dioxide and used as a propellant.

Figure 6.32 Two molecules rearrange to form one larger molecule.

An equilibrium is set up in this reaction.

According to Le Chatelier's principle, if the **concentration** of one component is altered the equilibrium will respond to counteract the change.

If $N_2O_4(g)$ is removed as it is formed, the concentration of it decreases so the equilibrium moves to the *right*, as more is made to replace it.

If more NO_2 (the reactant) is introduced, the concentration of it increases so again the equilibrium moves to the right, making more product N_2O_4.

When a product of a reaction is a gas and it is removed, this in effect reduces the concentration of the product. So the **position of the equilibrium** moves to the right.

KEY INFORMATION

Remember that the (g) following the formula is a state symbol, and stands for gas.

1 In the reaction

$$2SO_2(g) + O_2(g) \rightleftharpoons 2SO_3(g)$$

Predict which gas would need to be removed to make more product.

Changing concentrations

If the concentration of one of the reactants or products is changed, the system is no longer at equilibrium and the concentrations of all the substances will change until equilibrium is reached again.

- If the concentration of a reactant is increased, more products will be formed until equilibrium is reached again.
- If the concentration of a product is decreased, more reactants will react until equilibrium is reached again.

An increase in the amount of the product is the goal of manufacturers of bulk chemicals and so methods to increase the amount of product are important. Many of these reactions involve equilibrium reactions – one example is making ammonia.

The effect of reducing the concentration

When making ammonia in the Haber process, the reaction at equilibrium is:

$$N_2(g) + 3H_2(g) \rightleftharpoons 2NH_3(g)$$

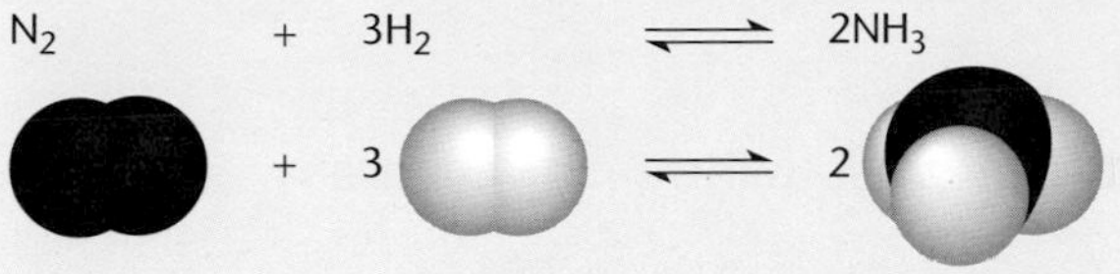

N_2 + $3H_2$ $\rightleftharpoons$ $2NH_3$

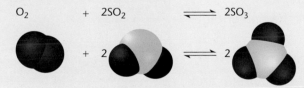

Figure 6.33 Four molecules rearrange to form two molecules.

If ammonia (NH_3) is liquefied and removed during the process more nitrogen and hydrogen (the reactants) will react to make more ammonia (the product) and attempt to restore equilibrium.

Sulfur trioxide (SO_3), is a starting chemical for the production of sulfuric acid to make fertilisers, paints and plastics.

The reaction at equilibrium is:

$$2SO_2(g) + O_2(g) \rightleftharpoons 2SO_3(g)$$

If sulfur trioxide, (SO_3), is removed during the process more sulfur dioxide and oxygen will react to produce more sulfur trioxide by restoring equilibrium.

O_2 + $2SO_2$ $\rightleftharpoons$ $2SO_3$

Figure 6.34 Three molecules rearrange to form two molecules.

2 Methanol is produced in industry by the reaction of hydrogen with carbon monoxide. Suggest how more methanol could be produced.

$$CO(g) + 2H_2(g) \rightleftharpoons CH_3OH(g)$$

3 Lead(II) chloride is a white solid. It is sparingly soluble in water and the following equilibrium is set up. Explain what you would observe if sodium chloride was added.

$$PbCl_2(s) \rightleftharpoons Pb^{2+}(aq) + 2Cl^-(aq)$$

Changing temperature and equilibrium

Learning objectives:

- explain how exothermic reactions behave
- explain how endothermic reactions behave
- apply Le Chatelier's principle to reactions in equilibrium.

Exothermic and endothermic reversible reactions behave in different ways when the temperature is increased or decreased. Applying Le Chatelier's principle to each case will explain how the yield of a product can be maximised.

HIGHER TIER ONLY

Changes with increasing temperature

The percentages of reactants and products for a reaction at equilibrium at different temperatures are shown in Table 6.4.

This reaction is: $A \rightleftharpoons B$

Temperature in °C	Percentage of *reactants* at equilibrium	Percentage of *products* at equilibrium
20	38	62
30	45	55
40	52	48
50	56	44

Table 6.4

You can see that as the temperature increases the percentage of products decreases.

Is this an **exothermic** reaction or an **endothermic** reaction?

Le Chatelier's principle states: if a system is at equilibrium and a change is made to any of the conditions, then the system responds to *counteract* the change.

If this principle is applied to this reaction, then the temperature increase is causing the reaction to go *backwards*.

To make the reaction go forwards (to produce more product) we can predict that the temperature must be lowered.

Why is this?

This will be because the forward reaction ($A \rightarrow B$) is transferring out energy.

As it is transferring out energy, any external increase in temperature (i.e. raising the temperature of the system) will need to be counteracted by the system as there is too much energy in the system. So the reaction goes backwards.

We can deduce then, that this forward reaction is an exothermic reaction.

The opposite is true for a forward reaction that is endothermic. The external temperature will need to be raised.

So, in summary for formation of products:

If the temperature of a system at equilibrium is *increased*:	If the temperature of a system at equilibrium is decreased:
the relative amount of products at equilibrium *increases* for an *endothermic* reaction	the relative amount of products at equilibrium *decreases* for an *endothermic* reaction
the relative amount of products at equilibrium *decreases* for an *exothermic reaction*.	the relative amount of products at equilibrium *increases* for an *exothermic* reaction.

1 Use Table 6.4 to predict the percentages of A and B in the equilibrium mixture at (a) 45 °C and (b) 55 °C.

2 Explain why an increase in temperature would favour the making of products in an endothermic reaction.

Applying Le Chatelier's principle

The process used to make sulfur dioxide into sulfur trioxide is an exothermic reaction.

$$2SO_2(g) + O_2(g) \rightleftharpoons 2SO_3(g)$$

This means that to increase the *yield* of products the temperature of the reaction needs to be lowered.

The graph shows that *lowering* the temperature increases the % conversion. It does this by altering the **position of the equilibrium** to the right to counteract the lowering temperature.

3 Use Figure 6.35 to explain why the process is endothermic in the backward reaction.

4 A substance F is in equilibrium with substance E.

$$E(g) \rightleftharpoons 2F(g)$$

Some data is given in the table.

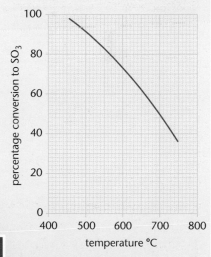

Figure 6.35 Graph showing % sulfur trioxide conversion from SO_2

	% Conversion to F at different temperatures and pressures		
	1 atmosphere	5 atmospheres	10 atmospheres
20°C	15%	10%	2%
100°C	35%	24%	4%

a Explain the variation in % conversion with temperature.

b Justify whether the reaction is exothermic or endothermic.

c Using Le Chatelier's principle, suggest why % conversion changes with pressure.

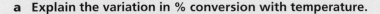

Changing pressure and equilibrium

Learning objectives:

- predict the effects of changes in pressure
- explain why these effects occur
- interpret data to predict the effect of a change in pressure.

The yield in industrial processes involving gases can be increased by changing the pressure, according to Le Chatelier's principle. The yield of product in all the reactions between gases can be maximised by making changes in all three conditions – concentration, temperature and pressure – but sometimes compromises have to be made.

HIGHER TIER ONLY

Pressure change and yield

When a reversible reaction takes place and reaches equilibrium, the position of the equilibrium shifts if the conditions are changed.

The data in Table 6.5 shows the percentages of reactants and products for a reaction through a range of **pressures**.

Pressure in atmospheres	Percentage of reactants at equilibrium	Percentage of products at equilibrium
1	33	67
5	41	59
10	55	45
15	62	38

Table 6.5

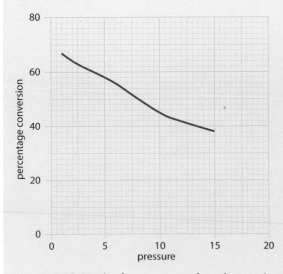

Figure 6.36 Graph of percentage of product against pressure

1. Predict the percentage equilibrium composition of both reactants and products at 7.5 atmospheres.

2. Describe the trend in the percentage of products with the change in pressure.

Examples of pressure change in processes

The reaction between nitrogen and hydrogen to form ammonia is reversible:

$$N_2(g) + 3H_2(g) \rightleftharpoons 2NH_3(g)$$

The number of molecules as reactants is 4 (3 + 1)	The number of molecules as products is 2

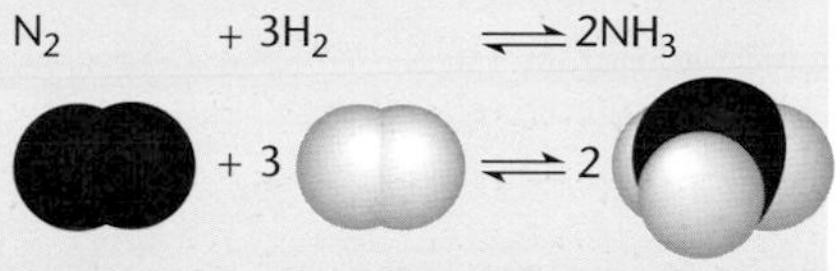

$N_2 \quad + 3H_2 \quad \rightleftharpoons 2NH_3$

Figure 6.37 Four molecules of reactant become two molecules of product.

In summary for gaseous reactions at equilibrium:

- an *increase* in pressure causes the **equilibrium position** to shift towards the side with the *smaller* number of molecules as shown by the symbol equation for that reaction
- a decrease in pressure causes the equilibrium position to shift towards the side with the *larger* number of **molecules** as shown by the symbol equation for that reaction.

3. Explain the change in pressure needed to increase the percentage of product in these gaseous reactions.

    a $2A + 3B \rightleftharpoons 2C$

    b $A + B \rightleftharpoons 3C$

    c $A + B \rightleftharpoons 2C + D$

Reaching a compromise

For each process the conditions that can be changed are concentration, temperature and pressure. In practice all three have to be considered together and a **compromise** may be needed.

When there are four gas molecules on the left of the equation and two on the right, high pressure forces the equilibrium further to the right which increases the yield.

Sometimes the temperature used needs to be a compromise. If the forward reaction is exothermic, a high temperature reduces the yield and drives the equilibrium to the left to make more reactant. However, high temperatures increase the rate of reaction, so the chemical is produced faster.

4. Compare and explain the similarities and differences between the equilibrium conditions used in two processes, X and Y. Suggest with reasons whether each reaction is endothermic or exothermic and whether there are more reactant molecules than product molecules or vice versa.

	Process X	Process Y
Temperature / °C	400	15
Pressure / atmospheres	200	1–2

KEY INFORMATION

The changes that happen follow Le Chatelier's principle: if a system is at equilibrium and a change is made to any of the conditions, then the system responds to counteract or reverse the change.

DID YOU KNOW?

A catalyst is a small amount of a chemical that speeds up a particular reaction, without being used up. It does not alter the equilibrium position or the yield of product but makes it feasible or produces more every second.

MATHS SKILLS

Use the slope of a tangent as a measure of rate of change

Learning objectives:

- draw graphs from numeric data
- draw tangents to the curve to observe how the slope changes
- calculate the slope of the tangent to identify the rate of reaction.

KEY WORDS

tangent
rate
slope
graph

We have found that reactions go faster or slower by changing the temperature, concentration of a solution or the size of solid pieces. Can you read graphs well to see the effects by looking at the slopes of the lines? Can you calculate the slopes of those lines? Can you draw tangents at any point of the reaction lines to calculate the rate?

Drawing graphs from data

When you have investigated the reaction between magnesium and acid you will have collected hydrogen. You will have measured the volume of hydrogen given off every 10 seconds for 60 seconds.

This is an example of the data you would collect:

Time in secs	0	10	20	30	40	50	60
Volume of H_2 in cm³	0	22	42	59	66	66	66

1 Draw a graph of the data in the table. Mark where the reaction is:
   a fastest **b** slowing down **c** stopped.

2 Explain in terms of collision theory when the rate of reaction is:
   a highest **b** zero.

Finding the gradient

To find the gradient, m, of a straight line, you need to find the coordinates of two points, (x_0, y_0) and (x_1, y_1). The gradient of the line is the change in y (the vertical distance) compared with the change in x (the horizontal distance). This is calculated by dividing $(y_1 - y_0)$ by $(x_1 - x_0)$.

So $m = (y_1 - y_0) \div (x_1 - x_0)$.

Measure the value of y up the line (here it is volume of H_2) and the value is 50.0 cm³.

Then measure the value of x along the axis (here it is time) and the value is 25 secs.

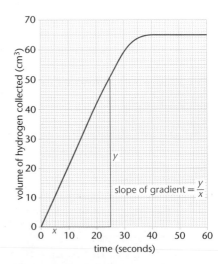

Figure 6.38 Calculating gradients

To find the gradient, divide y by x. So in this graph:

$m = (y_1 - y_0) \div (x_1 - x_0)$

$m = (50 - 0) \div (25 - 0) = \dfrac{50}{25} = 2$

Determine the units from those given on the axes so

$\dfrac{y}{x} = \dfrac{50}{25} = 2$ cm³/sec.

Quantity of reactant or product	Unit of quantity	Unit of rate
Mass	g	g/s
Volume	cm³	cm³/s
Moles	mol	mol/s

3 Look at Figure 6.38.
  a Work out the volume of gas that was collected between 10 and 30 seconds.
  b Calculate the rate of reaction during this 20 second interval.
  c Calculate the gradient between 0 and 10 seconds.
  d State why the gradient from part (b) is different from that calculated in part (c).

Drawing tangents to a curve

It is usually necessary to draw two graphs to compare the rate of a reaction when factors have changed. The results might look like Figure 6.39.

Gradients can be calculated for both lines and the rates can be compared.

Look at the blue line. There is some slowing down in the reaction. To measure the rate of reaction during this period a tangent to the curve needs to be drawn. Choose a point such as 25 secs and put your ruler so that it is at the best position to follow the curve direction *at that point*. Lengthen this tangent line so that it hits convenient major grid-lines. Draw a triangle using the tangent line as the hypotenuse. Measure the value of y and the value of x. Calculate the gradient $\frac{y}{x}$ at that point.

4 Draw construction lines to calculate the gradients of the blue and red lines of Figure 6.39. What is the rate of reaction calculated from:
  a the blue line?
  b the red line? Comment on your findings.

5 On Figure 6.39 draw tangents to the blue curve at 22 seconds and 28 seconds. Calculate the rate of reaction at both points and comment on your findings.

6 Rate data from a reaction between a carbonate and an acid was plotted. Two tangents were taken from a volume–time graph. They were 2.7 cm³/minute and 8.5 cm³/minute.
  a Explain which tangent was taken closer to the end of the reaction.
  b Work out the volume of gas produced in 1 second using the tangent taken nearer the start of the reaction.

DID YOU KNOW?

Sometimes it is easier to measure the *mass* of a product formed. The rate of reaction is then measured in g/s or g/min.

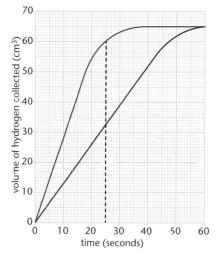

Figure 6.39 Different gradients

REMEMBER!

It is very important to check the units you are using.

Check your progress

You should be able to:

identify how to measure the amount of gas given off in a reaction →	calculate the mean rate of a reaction →	draw tangents to the curves as a measure of the rate of reaction
identify which factors affect the rate of reactions →	explain how changes of surface area affect rates →	explain how rates are affected by different factors
analyse experimental data on rates of reaction →	predict the effects of changing conditions on rates of reactions →	explain the effects of changes of a factor on rates of reaction using collision theory
identify catalysts in reactions →	explain catalytic action →	explain activation energy
identify a reversible reaction →	describe how equilibrium is reached →	predict the effects of changes on systems at equilibrium
identify reactants and products in a reversible reaction →	explain how changing concentrations changes equilibrium →	interpret data to predict the effect of a change in concentration
explain how exothermic reactions behave if the temperature of systems at equilibrium changes →	explain how endothermic reactions behave if the temperature changes →	interpret appropriate data to predict the effect of a change in temperature on reactions at equilibrium
predict the effects of changes in pressure →	explain why these effects of pressure change occur →	apply Le Chatelier's principle to reactions in equilibrium

Worked example

1 **Which of these reactions is the fastest?**

cement setting   rusting   (acid on magnesium pieces)

> The answer acid on magnesium pieces is correct

2 **Draw an energy profile to show the activation energy needed for a reaction to take place.**

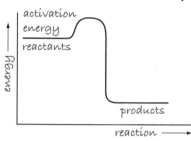

> The profile is for an exothermic reaction. The activation energy is the *difference between* the energy levels of the reactants and the 'hump', not the initial energy level.

3 **Explain how catalysts work.**

They speed up a reaction because they lower the activation energy.

> This answer is correct, as the description of what catalysts do is linked to the explanation.

4 **Explain how increasing temperature increases the rate of reaction.**

The particles move faster so hit each other more likely.

> This answer is worth one mark as a better word would be collide and an explanation is needed. A better answer is that the particles move with more energy so there are more successful collisions.

5 **Draw the equation symbol used to show a reversible reaction.**

> The answer is correct.

6 **Look at the table. Fill in the missing number and describe the pattern shown by this reaction at equilibrium. Predict the percentage of product probably formed at 150 °C and suggest what the owner of the factory making this product should do.**

Temperature in °C	100	200	300	400
% reactants at equilibrium	22	37	52	59
% products at equilibrium	78	63	48	41

The higher the temperature the less products are made, so the owner should do this at a low temperature.

> The data on the table is completed and the pattern is correctly described. The student has made a good suggestion to use a low temperature but needs to be careful to use the word lower not low. This reaction may not work at a low temperature and may not be fast enough. This time the 'benefit of doubt' is given.
>
> The answer is not complete though as the student has missed predicting the percentage of product at 150 °C. The prediction should have been between 70%–71%, a value between the percentages at 100 °C and 200 °C

End of chapter questions

Getting started

1. What unit is the volume of a gas measured in? Circle your answer.

   g cm^2 cm^3

   `1 Mark`

2. Which piece of apparatus is used to measure gas volume?

   a top pan balance   **b** glass syringe   **c** plastic pipette

   `1 Mark`

3. Write down two factors that can be changed to make a reaction go faster.

   `2 Marks`

4. Two solutions are mixed together. The reaction goes faster if a small amount of a solid is added. What is this solid acting as?

   a catalyst   **b** product   **c** indicator

   `1 Mark`

5. If 1 g magnesium is added to acid each time, first as ribbon, then as ribbon pieces, then as powder, which will make the fastest reaction?

   `1 Mark`

6. Acid is added to sodium carbonate in a flask and left on a digital balance. Explain why the mass of flask and contents goes down.

   `2 Marks`

7. Match the result readings to the experiment.

sec	15 27 35 42 57

g	104.8 103.5 102.8

cm^3	0 20 25 32 39

   rate of reaction by loss of mass

   rate of reaction by obscuring with a deposit

   rate of reaction by gas volume

   `2 Marks`

Going further

8. What is the symbol for a reversible reaction?

   `1 Mark`

9. What happens to the rate of a gas reaction when the pressure is increased?

   a stays the same   **b** increases   **c** decreases

   `1 Mark`

10. Describe the role of a catalyst.

    `2 Marks`

11. Describe *one* experiment you would do to show that the rate of reaction between zinc and hydrochloric acid (to make hydrogen) can go faster.

    `4 Marks`

12. Sam and Alex think that when acid is added to marble chips the reaction is slower than when the same acid is added to marble powder. Explain two factors they should keep the same when they check if their results fairly compare the rate of the two reactions they are testing.

    `2 Marks`

More challenging

13. Describe activation energy.

    `1 Mark`

14. What kind of reaction has both a forward and a backward reaction?

    `1 Mark`

15. Describe the features of a reaction in equilibrium.

    `2 Marks`

16. In the reaction $N_2 + 3H_2 \rightleftharpoons 2NH_3$ increasing the pressure increases the amount of product. Explain why more ammonia will be made.

    `2 Marks`

More demanding

17 In an experiment a student reacted acid with potassium carbonate. The carbon dioxide was collected in a gas syringe. These are the results.

Temperature/°C	Volume of gas collected in 10 seconds/cm³		
18	30	32	31
23	40	45	44
28	59	60	58
33	73	76	72

Explain which variable was being investigated, which was being measured and identify the result which did not follow the pattern. Describe the trend in the results.

4 Marks

18 Explain the reason why the temperature may be lowered for a reaction in equilibrium.

2 Marks

19 The conditions for making chemical X are given below. The percentage of X achieved at each pair of conditions is given in the table.

Pressure/atm Temperature/°C	200	300	400	500
350	43	56	64	71
400	32	36	49	58
450	21	25	35	40
500	15	18	22	28

a The making of chemical X from two gases is affected by temperature and pressure. Suggest what kind of reaction this is.

b Explain why a low temperature and high pressure are chosen, using trends seen in the data in the table.

c Suggest why a temperature lower than 350 °C is not chosen.

4 Marks

20 Use the data in the table in question 17 to evaluate this claim 'For every 10 °C rise in temperature the rate of reaction doubles'. Explain how you would calculate the rate of reaction at 28 °C using a graph of these results.

4 Marks

Total: 40 Marks

HYDROCARBONS

CRUDE OIL AND HYDROCARBONS

• Crude oil is a fossil fuel made millions of years ago.
• We get petrol and diesel from crude oil.
• Petrol is easier to ignite than coal.

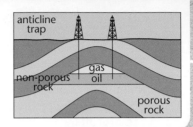

PROPERTIES OF HYDROCARBONS

• Bunsen burners provide either a blue flame or a yellow flame.
• We need fuels for heating and transport.
• Carbon atoms join with hydrogen atoms to make molecules.

COMBUSTION

• Natural gas burns with a blue flame when there is plenty of oxygen.
• Natural gas burns with a yellow flame when there is a shortage of oxygen.
• Carbon dioxide turns limewater milky.

FUEL SUPPLY AND DEMAND

• We need petrol and diesel for car transport.
• We drill for crude oil on land and under the sea.
• Crude oil is a finite resource.

OTHER USES OF CRUDE OIL

• Crude oil is separated by distillation.
• Some parts of crude oil are processed to make plastics.
• Some parts of crude oil are processed to make detergents.

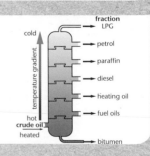

IN THIS CHAPTER YOU WILL FIND OUT ABOUT:

CRUDE OIL AND HYDROCARBONS

- How is crude oil separated?
- What are the hydrocarbons that make up crude oil?
- How does the size of hydrocarbons affect flammability?

PROPERTIES OF HYDROCARBONS

- What is incomplete combustion?
- How is the supply of petrol increased to match the demand?

WHAT ARE THE PRODUCTS OF COMBUSTION?

- Hydrocarbons burn to make carbon dioxide and water.
- Hydrocarbons burning in a shortage of oxygen make carbon monoxide.
- Carbon monoxide is highly toxic.

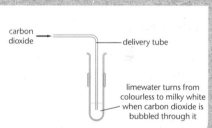

carbon dioxide

delivery tube

limewater turns from colourless to milky white when carbon dioxide is bubbled through it

HOW DOES CRACKING INCREASE SUPPLY?

- There is less demand for large hydrocarbon molecules.
- Large hydrocarbon molecules are cracked to make smaller molecules.
- Smaller molecules in petrol are in high demand.

WHAT ARE ALKENES?

- Alkenes made during cracking are smaller hydrocarbons.
- Alkenes are more reactive hydrocarbons.
- Alkenes decolourise bromine water.

Crude oil, hydrocarbons and alkanes

Learning objectives:

- describe why crude oil is a finite resource
- identify the hydrocarbons in the series of alkanes
- explain the structures and formulae of alkanes.

KEY WORDS

alkane
crude oil
finite resource
hydrocarbon

Crude oil is a fossil fuel made millions of years ago. When it runs out there will be no more. This leads to lots of questions. What do we use it for now? How much of it do we use? When will it run out? What will happen when it does run out? How wisely do we use it?

Crude oil

Crude oil is a fossil fuel. It was made millions of years ago. Fossil fuels are **finite resources** because they are no longer being replaced. The conditions on Earth are not the same as they were millions of years ago. They are called non-renewable sources, as they cannot be made again.

Figure 7.1 Drilling for crude oil

DID YOU KNOW?

There is a theory called 'peak oil' that talks about the oil running out and what will happen. You can look this up to find out more.

Crude oil is found in rocks. It was made from the remains of plankton and other ancient biomass compressed in mud over millions of years. It is made up of a mixture of many types of oils. All these oils are **hydrocarbons**. Hydrocarbons are made of carbon and hydrogen bonded together.

1. Name one other fuel that is a fossil fuel.

2. Suggest why it is important we use less fossil fuels.

Hydrocarbons

Hydrocarbons are molecules made from carbon and hydrogen *only*. The smallest hydrocarbon is called methane. One atom of carbon and four atoms of hydrogen chemically combine to make it.

Methane is an **alkane**. Alkanes are one series of hydrocarbons. You will meet another series later.

The first four hydrocarbons in the alkane series are methane, ethane, propane and butane.

(a)

$$H-\overset{\overset{\displaystyle H}{|}}{\underset{\underset{\displaystyle H}{|}}{C}}-H$$

(b)

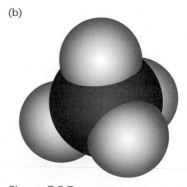

Figure 7.2 Two ways to represent the molecule of methane CH_4

Name	Number of carbon atoms	Formula	Displayed formula
Methane	1	CH_4	$$\begin{array}{c} H \\ | \\ H-C-H \\ | \\ H \end{array}$$
Ethane	2	C_2H_6	$$\begin{array}{cc} H & H \\ | & | \\ H-C-C-H \\ | & | \\ H & H \end{array}$$
Propane	3	C_3H_8	$$\begin{array}{ccc} H & H & H \\ | & | & | \\ H-C-C-C-H \\ | & | & | \\ H & H & H \end{array}$$
Butane	4	C_4H_{10}	$$\begin{array}{cccc} H & H & H & H \\ | & | & | & | \\ H-C-C-C-C-H \\ | & | & | & | \\ H & H & H & H \end{array}$$

3 Explain why the formula for the next one in the series is C_5H_{12}

4 Predict the formula for the alkane that has eight carbon atoms.

Homologous series

All the alkanes have the same general formula, so they form one series of compounds. The general formula is C_nH_{2n+2}

One carbon atom forms four bonds. Hydrogen has one bond, so four hydrogen atoms bond to one carbon atom to make methane.

All the bonds in alkanes are single bonds. If one carbon bonds to another carbon atom to make ethane, the carbon–carbon bond is a single bond. Each carbon atom has three bonds left to make bonds with hydrogen atoms. The formula can be written as CH_3CH_3

Look back at the displayed formula of butane in the table. How many hydrogen atoms are bonded:

- to each of the carbon atoms at either end?
- to each of the carbon atoms in the middle?

You can also write the formula for butane as $CH_3CH_2CH_2CH_3$

5 Explain how to use the general formula to work out the formula of a hydrocarbon with 18 carbon atoms.

6 Write the formula C_7H_{16} in two other ways, remembering to only use single bonds.

7 Alkanes can also form branched chains. Draw the displayed formula for the straight-chain alkane and **two** branched-chain alkanes with the formula C_5H_{12}

Fractional distillation and petrochemicals

Learning objectives:

- describe how crude oil is used to provide modern materials
- explain how crude oil is separated by fractional distillation
- explain why the boiling points of the fractions are different.

KEY WORDS

boiling point
fractional
  distillation
hydrocarbon
liquefied
  petroleum gas

Crude oil is a smelly, yellow-to-black liquid that we can change to provide us both with fuels we use every day and the chemicals needed to make synthetic fabrics, medicines and dyes. How do we do this? The mixture has to be separated by boiling and collecting the different parts.

Products from crude oil

The mixture of oils in crude oil is separated into 'fractions' by **fractional distillation**. Then each fraction can be processed to produce fuels and feedstocks for the petrochemical industry.

Many of the useful materials on which we depend for our modern lifestyle are produced from crude oil:

- fuels, such as **liquefied petroleum gases (LPG)**, petrol, kerosene, diesel oil, heavy fuel oil and
- petrochemicals, such as solvents, lubricants, polymers and detergents.

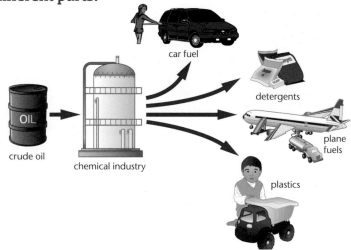

Figure 7.3 Useful products from crude oil

The huge number of natural and synthetic carbon compounds occurs because of the ability of carbon atoms to form 'families' of similar compounds using its four bonds.

1 **Explain the difference between a fuel and a petrochemical.**

Separating fractions

Fractional distillation is the process of separating the mixture of **hydrocarbons** in crude oil. The oil is evaporated and allowed to cool and condense. The different fractions have different **boiling points** and so will condense at different temperatures.

Crude oil is heated at the bottom of a tower, so it is hot when it enters the bottom of the fractionating column (see Figure 7.4). The top of the tower is colder. The column has a temperature gradient, hotter at the bottom, cooler at the top.

- Most fractions boil and their gases rise up the tower towards the colder top.
- Fractions such as liquid petroleum gas (LPG) have low boiling points. They stay as a gas and exit at the top of the tower.

KEY INFORMATION

Liquids boil at their boiling point and the hot gases condense back to liquids at *that same temperature*. When liquid petrol has turned to hot petrol vapour it travels up the tower at a temperature higher than its boiling point. When it reaches a colder part of the tower that is the same temperature as the boiling point of petrol, the vapour will condense back to liquid.

- Fractions with slightly higher boiling points, such as petrol and kerosene, boil and stay as a gas until they reach a point in the tower where it is cold enough that they condense.
- Fractions with high boiling points, such as heavy fuel oils, boil but then condense first and exit near the bottom of the tower.
- Bitumen has such a high boiling point that it does not boil at all and sinks as a thick liquid to the bottom of the tower. This fraction is used to make tar for road surfaces.

Each fraction from the distillation contains a mixture of hydrocarbons with similar boiling points. Each fraction contains molecules with a similar number of carbon atoms.

2   **Explain why crude oil itself does not have any uses.**

3   **Suggest which of the following hydrocarbons are likely to be in the same fraction:**

C_2H_6 C_6H_{14} C_7H_{16} $C_{11}H_{24}$ $C_{20}H_{42}$

Justify your answer.

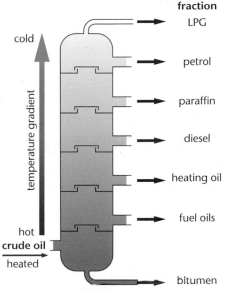

Figure 7.4 A fractional distillation column

HIGHER TIER ONLY

Why can fractions be separated?

Crude oil can be separated by fractional distillation because the molecules in different fractions have different numbers of carbon atoms and so have different length chains. This means that the total quantity of the weak forces *between* the molecules is different.

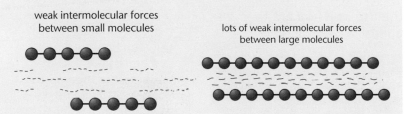

Figure 7.5 Forces *between* molecules. Large molecules such as bitumen have more weak forces of attraction between them (shown by continuous wavy lines). Smaller molecules have fewer weak forces between them (shown by short wavy lines).

These weak forces *between* the molecules are broken during boiling. When a liquid boils, the molecules separate from each other as molecules of gas.

Large molecules, such as those that make up bitumen and heavy oil, have very long chains so there are lots of weak forces of attraction between the molecules along the chains. This means that they are difficult to separate. A lot of energy is needed to pull each molecule away from another. They have high boiling points.

4   **Hydrocarbon X condenses at 68 °C and hydrocarbon Y at 98 °C. Explain why.**

KEY INFORMATION

Remember that the hot chemicals enter the tower at the bottom. You may get a clearer understanding of this process if you read this part of the text again, starting at the bottom.

DID YOU KNOW?

Bonds joining atoms *in* a molecule are much stronger than forces *between* molecules. Bonds joining carbon and hydrogen atoms in a hydrocarbon molecule do not break when the substance is boiling.

Properties of hydrocarbons

Learning objectives:

- describe how different hydrocarbon fuels have different properties
- identify the properties that influence the use of fuels
- explain how the properties are related to the size of molecules.

KEY WORDS

flammability
ignite
viscosity
volatile

Fuels burn in oxygen to give us the energy we need for keeping warm and moving around. We need many different types of fuels for our different purposes. Each type of fuel has different properties that influence the way we use it.

Using fuels

Fractional distillation of crude oil results in many types of hydrocarbon fuels that have different properties. Their properties influence *how* they are used as fuels.

Petrol and diesel are used as fuels in car engines. Petrol and diesel are both flammable so they both burn to release energy. They are both liquids so are transported more easily than a gas.

Petrol flows easily and has a fairly low boiling point so it vaporises easily. This means it ignites easily. Diesel is more viscous and has a higher boiling point so it is less **volatile**. It **ignites** less easily than petrol.

The properties that are related to the use of a fuel are their boiling point, **viscosity** and **flammability**.

Figure 7.6 Different uses need different fuels

DID YOU KNOW?

Diesel has a higher freezing point than petrol. This is why diesel cars and lorries can have trouble starting in very cold winters. As the temperature falls, diesel can stop being a liquid at a higher (i.e. less cold) temperature than petrol. You could have problems with a diesel engine at around −15 °C but a petrol engine would still work well at −40 °C.

Fuel	Number of carbon atoms	Boiling point °C	Viscosity Relative scale 1 low – 10 high	Flammability Relative scale 1 poor – 10 good	Use
LPG	3–4	Below 30	Gas	10	Camping stoves
Petrol	6–10	80–120	2	8	Cars
Diesel oil	9–16	110–190	3	6	Cars and lorries
Kerosene	12–19	130–230	6	5	Aircraft
Heavy fuel oil	20–27	240–340	8	3	Ships

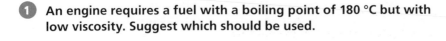

1 An engine requires a fuel with a boiling point of 180 °C but with low viscosity. Suggest which should be used.

Molecular size

The boiling point, viscosity and flammability of hydrocarbons depend on the size of their molecules. Look at the table. Petrol has 6–10 carbon atoms in its molecules. Diesel has 9–16 carbon atoms so its molecules are longer.

The boiling point of petrol is lower than the boiling point of diesel. So from these two data points we could say that the larger the molecule is the higher the boiling point.

Two data points is not enough for a trend. Look at all the data from the two columns. Is the predicted trend correct? Yes – from five data points we can now say the larger the molecule is the higher the boiling point.

2 **Use data from the table to describe the trends in:**

   a **viscosity**

   b **flammability**

   as the number of carbon atoms increases.

3 **Predict the boiling point of the fraction of hydrocarbons with 28–32 carbon atoms.**

Why trends in properties occur

You have seen from the table that the boiling point, viscosity and flammability of hydrocarbons depend on the size of their molecules.

Boiling point and viscosity *increase* with increasing molecular size. Flammability *decreases* with increasing molecular size.

Many of the hydrocarbons in crude oil fractions are long chains. The larger the hydrocarbon chains the greater the number of the weak forces *between* the molecules, so more energy is needed to separate them. This means that more energy is needed for the molecules to be free enough to form a vapour.

So the *larger* the molecule is the *higher* the boiling point.

4 **Explain why the larger the hydrocarbon molecule is the more viscous the liquid. Use ideas about forces between molecules.**

5 **The graph below shows the melting points of alkanes.**

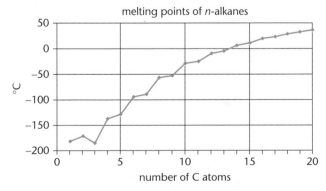

melting points of *n*-alkanes

°C — number of C atoms

   a **Explain the general trend in melting points.**

   b **The graph is not smooth. Identify a pattern between the number of carbons and melting point that takes this into account.**

(a)

(b)

Figure 7.7 Models of one of the molecules found in (a) petrol (b) diesel. Molecule (a) is shorter than molecule (b).

KEY INFORMATION

Forces *between* molecules are called *inter*molecular forces. You could remember this by thinking of *inter*national sporting matches which are matches *between* nations.

lots of weak intermolecular forces between large molecules

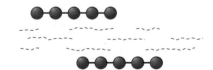

weak intermolecular forces between small molecules

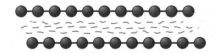

Figure 7.8 Larger molecules have stronger forces of attraction *between* them than the forces between smaller molecules.

Combustion

Learning objectives:

- describe the process of complete combustion
- balance equations of combustion of hydrocarbons
- explain the consequences of incomplete combustion.

Although coal is still burned to produce electricity, hydrocarbon fuels overtook the use of coal decades ago for many other purposes. Hydrocarbons have become indispensable to modern lifestyles as fuels for cars, lorries, trains, ships and aircraft and are also used in heating systems and manufacturing. They are cleaner than coal but still need to be burned correctly.

Burning hydrocarbon fuels

If a hydrocarbon fuel burns completely, carbon dioxide (CO_2) and water (H_2O) are made.

Complete **combustion** occurs when a fuel burns completely in air.

A fuel such as methane uses *oxygen* in the air to produce products. The products are **carbon dioxide** and water.

methane + oxygen → carbon dioxide + water

Incomplete combustion happens when a fuel burns in a shortage of oxygen.

More energy is released during complete combustion than during incomplete combustion.

Figure 7.9 Which flame shows that the methane fuel is burning completely?

DID YOU KNOW?

....

When fuel is burned in a shortage of oxygen, there is incomplete combustion and carbon dioxide is not the only product. Carbon monoxide is made, which can be fatal. Carbon monoxide detectors can be fitted near boilers in homes.

1. When the air hole on a Bunsen burner is open, the natural gas burns with a blue flame. When the air hole is closed, the flame is yellow. Explain why.

Complete combustion equations

Complete combustion releases useful energy, which is the purpose of a good fuel.

Methane is a common hydrocarbon fuel. Its formula is CH_4.

Complete combustion can start to be shown by the equation:

$$CH_4 + O_2 \rightarrow CO_2 + H_2O \qquad \text{(unbalanced)}$$

The equation must be made to balance.

In the reactants there are *two* oxygen atoms and there are four hydrogen atoms.	In the products there are three oxygen atoms and there are two hydrogen atoms.

The balanced equation is:

$$CH_4 + 2O_2 \rightarrow CO_2 + 2H_2O$$

In the reactants there are now *four* oxygen atoms and four hydrogen atoms.	In the products there are now *four* oxygen atoms and *four* hydrogen atoms.

The equation is now balanced.

During combustion, the carbon and hydrogen in the fuels are *oxidised*.

From the hydrocarbon, the carbon and the hydrogen both combine with oxygen. Carbon is oxidised to make carbon dioxide (CO_2) and hydrogen is oxidised to make water (H_2O).

KEY INFORMATION

To balance equations, first, count up the numbers of atoms *within* each molecule (shown by the *subscript* numbers). Don't change these numbers.

Then, if necessary, alter the number of molecules (the large number in *front* of the formula) in order to balance the numbers on either side of the arrow.

2 Work out the formula of the hydrocarbon that produces $4CO_2$ and $5H_2O$.

3 Write the balanced equation for the complete combustion of propane, C_3H_8.

Incomplete combustion

If hydrocarbon fuels do not have enough oxygen in which to burn then *incomplete combustion* takes place. Not only is less energy produced from the fuel but the products are water, **carbon monoxide** and carbon. Carbon is noticed as soot and the flame burns yellow as soot is also burning. Carbon monoxide is a highly toxic gas that has no smell. It can cause illness and can be fatal as the molecule competes with oxygen molecules to attach to the haemoglobin molecules in red bloods cells. It can attach about 200 times better than oxygen and when blood is circulated there is no oxygen delivered to the body. It is really important to have good ventilation when using appliances that burn hydrocarbon fuels.

4 Write a balanced equation for the incomplete combustion of decane, $C_{10}H_{22}$.

5 Octane, C_8H_{18}, can be completely or incompletely combusted.

   a Write balanced equations for the complete and incomplete combustion of octane.

   b Suggest which releases the most energy.

Cracking and alkenes

Learning objectives:

- describe the usefulness of cracking
- balance chemical equations as examples of cracking
- explain how modern life depends on the uses of hydrocarbons.

KEY WORDS

alkene
bromine
catalyst
cracking
supply and
  demand

The fractional distillation of crude oil provides petrol and many other useful fractions containing hydrocarbons for our modern lifestyle. But too much kerosene is produced and not enough petrol. This can be solved by turning the kerosene and other large hydrocarbons into smaller molecules. This process is called cracking.

Supply and demand

More petrol is needed, but less kerosene is needed.

It is difficult for industry to match supply with the demand for petrol. The solution to the problem of **supply and demand** is **cracking**. Industrial cracking plants are built near to refineries so that the large hydrocarbons can be cracked to make smaller, more useful molecules. The supply of petrol can then be matched more closely with demand as more petrol is produced.

The table contains an example of the data that a cracking plant could collect.

Product	Supply in tonnes	Demand in tonnes
Petrol	100	300
Diesel	200	100
Kerosene	250	50

Figure 7.10 Cracking alkanes on an industrial scale

1 **Look at the table. At this cracking plant what do they need to do to match supply with demand?**

Cracking

Kerosene can be broken down or 'cracked' into petrol. This can be done both on an industrial scale and also in the laboratory.

To crack a liquid alkane you need a high temperature and a catalyst.

This process involves heating long-chain hydrocarbons to vaporise them. The vapours are either passed over a hot **catalyst** or mixed with steam and heated to a very high temperature so that thermal decomposition reactions then occur. These processes are called catalytic cracking or steam cracking.

KEY INFORMATION

Remember you have learned that thermal decomposition means breaking down by using thermal energy.

Large molecules are not so useful. Large hydrocarbon molecules can be broken down or 'cracked' to produce smaller, more useful molecules. So cracking makes more petrol, and some of the shorter alkane products of cracking are also useful as fuels. However, the **alkenes** that are also produced can be used to produce polymers and as starting materials for the production of many other chemicals.

2 State two conditions that are needed for cracking to take place.

Testing for products of cracking

The products of cracking include alkanes and another type of hydrocarbon called *alkenes*.

This can be represented by an equation such as:

$$C_{16}H_{34} \rightarrow C_8H_{18} + 2C_3H_6 + C_2H_4$$
long-chain alkane   shorter alkane   alkene   alkene

Bromine water is used to test for alkenes. It is an orange solution. When an alkene is added the orange solution turns colourless.

When an alkane is added to the bromine water it remains orange because an alkane does not react with bromine.

> **DID YOU KNOW?**
>
> Zeolites are now used as modern catalysts for cracking. They have an open 'cage-like' structure.

> **KEY INFORMATION**
>
> Bromine solution is used to test for a double bond in alkenes. The orange bromine solution turns colourless because the bromine has reacted with the alkene and formed a new compound. The new compound is a dibromo compound, which is colourless.

Figure 7.11 Two test tubes of bromine solution: one with an alkane added to it and one with an alkene added to it. Which tube had the alkene added to it?

Alk*enes* are more reactive than alk*anes* and react with bromine water, turning it from orange to colourless. The test tube on the right is the alkene, as it has reacted with the bromine water.

3 Bromine solution is used to test two hydrocarbons, pent*ane* and pent*ene*. Predict which one will turn the solution from orange to colourless.

4 Balance the cracking equation by working out the numbers **D**, **E**, **F** and **G**.

$$\textbf{D } C_{14}H_{30} \rightarrow \textbf{E } C_7H_{16} + \textbf{F } C_3H_6 + \textbf{G } C_2H_4$$

5 Write a balanced equation for the cracking of $C_{10}H_{22}$ that produces **two** products.

KEY CONCEPT

Intermolecular forces

Learning objectives:

- recognise the strong covalent bonds within molecules
- recognise the weak intermolecular forces between molecules
- describe the effects of weak intermolecular forces on properties of substances.

KEY WORDS

intermolecular
covalent
force of
   attraction
intramolecular

Forces between molecules

The weak forces between molecules are the cause of many differences in the way molecules and materials behave and of differences in their physical properties. The reason why methane is a gas and petrol is a liquid is, in part, due to these weak forces. The reason why poly(ethene) is a solid that can be made into stretchy film is also linked to weak forces between poly(ethene) molecules. These weak forces are **intermolecular** forces.

Bonds within molecules

Carbon can make four strong bonds. Carbon atoms can bond with other carbon atoms, with some other non-metals and also with a few metals.

Carbon atoms can bond by using single bonds, double bonds and triple bonds. For example:

C–C   C=C   C≡C   C–H   C–O
C=O   C–N   C–Cl

These **covalent** bonds are *very strong* and do not break easily. These bonds help to make compounds such as those shown in these formulae.

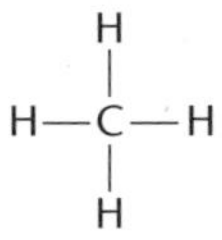

methane 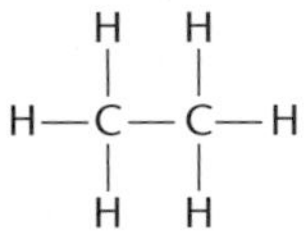   ethane 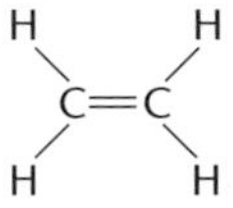   ethene 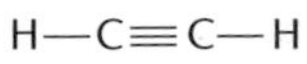   ethyne

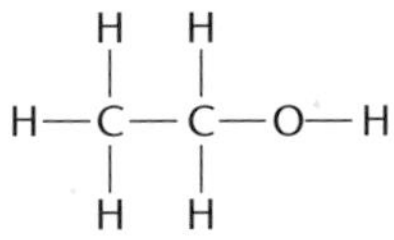

ethanol 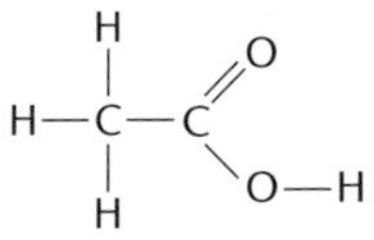   ethanoic acid 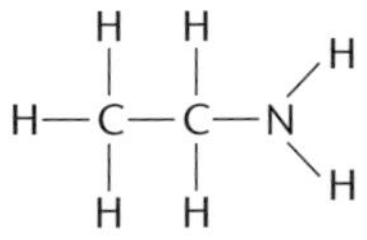   ethylamine

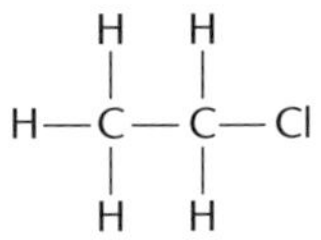

chloroethane 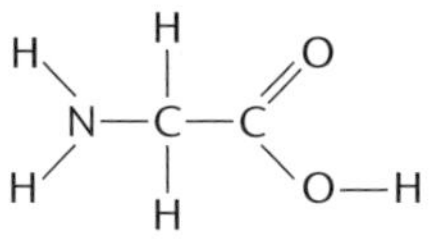   glycine 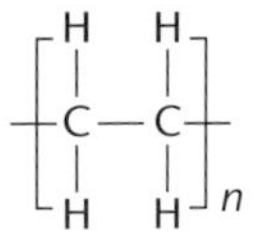   poly(ethene)

① Identify **two** molecules from those shown on the right that have double bonds.

Forces between molecules and changes of state

A molecule of methane does not break down easily. The four bonds *within* it are very strong. If two molecules of methane are side by side at room temperature there is a very weak **force of attraction** *between* the two molecules.

If there are lots of methane molecules there are more of the weak forces between them. The molecules move rapidly and randomly as a gas. When the temperature reduces to −161 °C the molecules have reduced their energy of movement and are closer together. The forces of attraction are greater so the gas condenses to a liquid.

If the liquid is heated up again, the molecules have more energy and the forces between them reduce once more so that the molecules can move freely as a gas again. They can continue to be heated but the molecules themselves will not break down. The bonds holding the atoms in each molecule together are so much stronger and do not break. These are strong bonds *within* molecules. These bonds are called **intramolecular** bonds.

The intermolecular forces between molecules are very weak. Those between methane molecules, CH_4, are weaker than those between octane molecules, C_8H_{18}. Octane is found in petrol.

Octane is a liquid at room temperature because there are more of the weak forces along the longer chains. Because there are more forces the longer chains need more energy to separate. They do not move from being part of a liquid to part of a gas as easily, so the temperature at which they can separate is higher, so the boiling point is higher.

2 Explain why decane has a higher boiling point than octane.

Intermolecular forces, stretching and rigidity

Methane molecules, CH_4 (found as a gas) are smaller than those of octane (found in liquid petrol), C_8H_{18}, but these are both much shorter than polymer chains such as poly(ethene) $-[CH_2CH_2]-_n$ which is a solid. The reason they are found at different states at room temperature is because the longer the molecule chains, the greater number of weak intermolecular forces between the molecules holding them together. So the long chains of polymer are held together as a solid.

The chains in poly(ethene) are only loosely held together so they can 'slide over' one another. This allows the material that is made up from the chains to stretch easily.

3 Propene has a melting point of −185 °C and poly(propene) a melting point of 68 °C. Explain why.

4 Melamine resin is used on worktop surfaces. Explain how this polymer will respond to a stretching force.

5 Ethene, $CH_2=CH_2$, has a boiling point of −104 °C. Ethane also has a two-carbon chain but has a boiling point of −89 °C. Explain why.

weak force of attraction between methane molecules

(a)

methane

(b)

octane

(c)

poly(ethene)

Diagrams of (a) methane molecules, CH_4, (b) octane molecules, C_8H_{18}, and (c) poly(ethene) $-[CH_2CH_2]-_n$ (where n can be greater than 10 000). Methane is a gas at room temperature, octane is a liquid and poly(ethene) is a solid.

weak intermolecular forces of attraction · pulling force · polymer stretches

Weak intermolecular forces; the polymer can stretch.

strong intermolecular chemical bonds or **cross-links** · pulling force · polymer is rigid

Strong intermolecular forces; the polymer cannot stretch.

MATHS SKILLS

Visualise and represent 3D models

Learning objectives:

- use 3D models to represent
  - › alkanes
  - › alkenes
  - › polymers.

KEY WORDS

3D model
representations
tetrahedron

We use carbon compounds every day as fuels, foods and materials. Carbon atoms are able to form a huge number of combinations with hydrogen, oxygen and other atoms to make up millions of different carbon compounds. We need to have a system of representing these compounds. Can you use 3D models to represent molecules of carbon compounds and then draw them?

Models of hydrocarbons (alkanes)

Carbon atoms are able to make four bonds, so can make whole series of similar compounds by adding another carbon atom to a chain. We use **3D models** to help us **represent** the compounds. They are especially useful as carbon atoms form four bonds in the shape of a **tetrahedron** (also called a triangular-based pyramid), so it is useful to see the direction of the bonds.

DID YOU KNOW?

The series of similar carbon compounds are called homologous series. Each carbon atom makes four bonds, but the bonds can be single, double or triple bonds.

REMEMBER!

Carbon always makes four bonds.

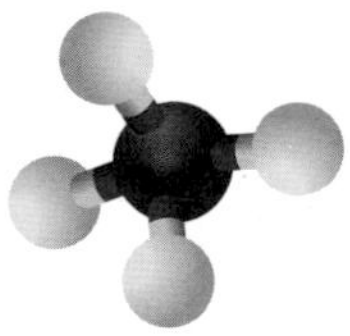

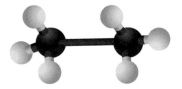

Figure 7.12 3D model of methane CH_4

Figure 7.13 3D model of ethane C_2H_6

These can be represented by 2D 'atom and bond' diagrams that show the type of bond but not the direction of it.

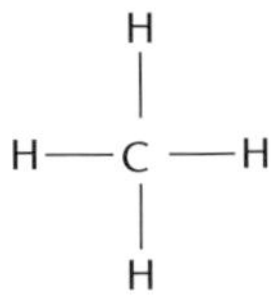

CH_4

C_2H_6

Figure 7.14 2D diagram of CH_4

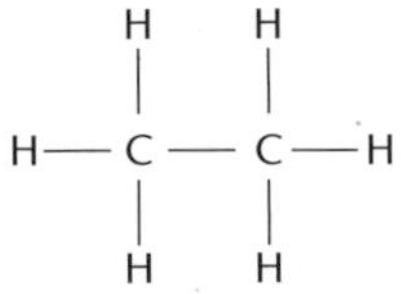

Figure 7.15 2D diagram of C_2H_6

If five carbon atoms join together, the compound in the same series as methane and ethane is known as pentane (the compound with eight carbon atoms is known as octane).

Figure 7.16 3D model of pentane C_5H_{12}

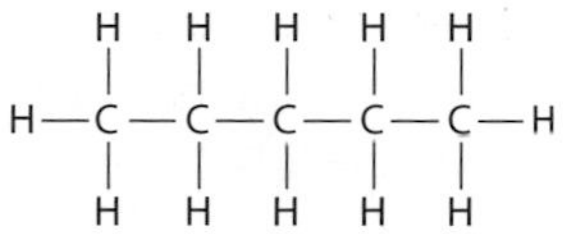

$$H-\overset{\overset{\displaystyle H}{|}}{\underset{\underset{\displaystyle H}{|}}{C}}-\overset{\overset{\displaystyle H}{|}}{\underset{\underset{\displaystyle H}{|}}{C}}-\overset{\overset{\displaystyle H}{|}}{\underset{\underset{\displaystyle H}{|}}{C}}-\overset{\overset{\displaystyle H}{|}}{\underset{\underset{\displaystyle H}{|}}{C}}-\overset{\overset{\displaystyle H}{|}}{\underset{\underset{\displaystyle H}{|}}{C}}-H$$

Figure 7.17 2D diagram of pentane C_5H_{12}

1 Draw a 2D diagram of a molecule of propane C_3H_8

2 What do these models represent?

a
$$H-\overset{\overset{\displaystyle H}{|}}{\underset{\underset{\displaystyle H}{|}}{C}}-\overset{\overset{\displaystyle H}{|}}{\underset{\underset{\displaystyle H}{|}}{C}}-\overset{\overset{\displaystyle H}{|}}{\underset{\underset{\displaystyle H}{|}}{C}}-\overset{\overset{\displaystyle H}{|}}{\underset{\underset{\displaystyle H}{|}}{C}}-\overset{\overset{\displaystyle H}{|}}{\underset{\underset{\displaystyle H}{|}}{C}}-\overset{\overset{\displaystyle H}{|}}{\underset{\underset{\displaystyle H}{|}}{C}}-H$$

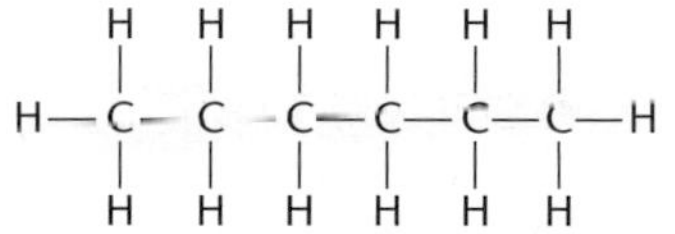

b
$$H-\overset{\overset{\displaystyle H}{|}}{\underset{\underset{\displaystyle H}{|}}{C}}-\overset{\overset{\displaystyle H}{|}}{\underset{\underset{\displaystyle H}{|}}{C}}-\overset{\overset{\displaystyle H}{|}}{\underset{\underset{\displaystyle H}{|}}{C}}-\overset{\overset{\displaystyle H}{|}}{\underset{\underset{\displaystyle H}{|}}{C}}-\overset{\overset{\displaystyle H}{|}}{\underset{\underset{\displaystyle H}{|}}{C}}-\overset{\overset{\displaystyle H}{|}}{\underset{\underset{\displaystyle H}{|}}{C}}-\overset{\overset{\displaystyle H}{|}}{\underset{\underset{\displaystyle H}{|}}{C}}-H$$

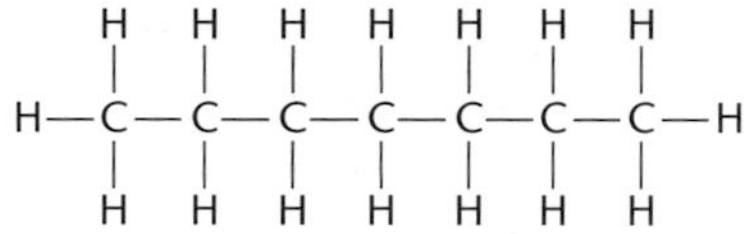

choose from hexane, heptane, decane or dodecane

Check your progress

You should be able to:

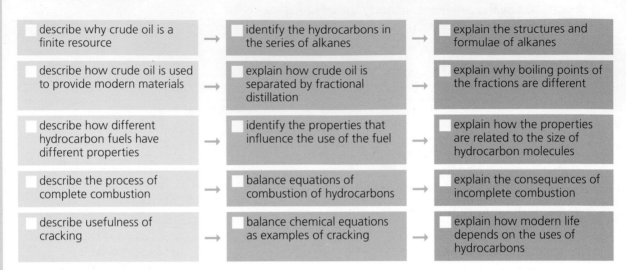

describe why crude oil is a finite resource → identify the hydrocarbons in the series of alkanes → explain the structures and formulae of alkanes

describe how crude oil is used to provide modern materials → explain how crude oil is separated by fractional distillation → explain why boiling points of the fractions are different

describe how different hydrocarbon fuels have different properties → identify the properties that influence the use of the fuel → explain how the properties are related to the size of hydrocarbon molecules

describe the process of complete combustion → balance equations of combustion of hydrocarbons → explain the consequences of incomplete combustion

describe usefulness of cracking → balance chemical equations as examples of cracking → explain how modern life depends on the uses of hydrocarbons

Worked example

1 **When propane is burned in a plentiful supply of air, carbon dioxide and water are made. Identify the gas made when there is a shortage of oxygen.**

carbon monoxide

> The answer is correct.

2 **Propane, C_3H_8, is one of a series of alkanes.**

Draw the full structural displayed formula of propane.

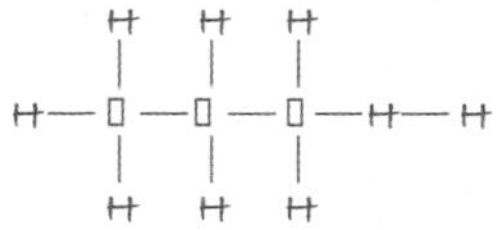

> The correct number of carbon atoms (3) have been drawn with the hydrogens on the carbons correctly attached, except there is one more H atom attached to another H atom. This makes 9 H atoms instead of 8. H makes one bond to C only, not to H.

3 **Propane is an alkane and propene is an alkene.**

Explain how you would test each to detect which one is propene.

Pass both of them through bromine water. The propane will not turn the bromine any other colour. The propene will turn the bromine another colour.

> Bromine water is the correct test reagent to use. The colour change is correctly caused by propene not propane. The colour change from orange / brown to colourless needs to be stated.

4 **Propene is an alkene. What can alkenes be used for?**

Making polymers

> The answer is correct.

5 **Explain the conditions needed for cracking large alkane molcules.**

High temperatures and a catalyst.

> The answer is correct.

6 **Write a balanced equation for the combustion of propane, C_3H_8, in a plentiful supply of oxygen.**

$$C_3H_8 + O_2 \rightarrow 3CO_2 + 4H_2O$$

> The reactants and product formulae are correct and the number of product molecules is correct. The oxygen atoms need to be balanced by putting $5O_2$ on the left-hand side.

End of chapter questions

Getting started

1 Heavy fuel oil exits near the bottom of the fractionating column during fractional distillation of crude oil. Which temperature range is this?

    a 80–120 °C   **b** 110–190 °C   **c** 130–230 °C   **d** 240–340 °C   `1 Mark`

2 Define the term hydrocarbon.   `1 Mark`

3 Identify the products for complete and incomplete combustion of an alkane.   `2 Marks`

4 Identify the alkane.

    a C_3H_8 **b** C_3H_6 **c** C_3H_8O **d** $C_3H_6O_2$ `1 Mark`

5 C_8H_{18} and $C_{20}H_{42}$ are both alkanes from crude oil. Predict which will have the higher boiling point.   `1 Mark`

6 $C_{30}H_{62}$ is a large alkane that needs 'cracking'. Explain what will happen when it is 'cracked'.   `2 Marks`

7 Match the process to the product   `2 Marks`

joining alkenes	carbon monoxide
incomplete combustion	carbon dioxide
complete combustion	polymer

8 State the name of the chemical used for testing alkenes.   `2 Marks`

9 Alkenes are used as the starting molecules for other chemicals. State one example of a type of molecule made from alkenes.   `1 Mark`

Going further

10 Explain the trend in boiling points of the alkane series.   `2 Marks`

11 Cracking is carried out on fractions produced from crude oil.

    a Explain the purpose of cracking.

    b Write a balanced equation for the cracking of heptane, C_7H_{16}.   `2 Marks`

12   `2 Marks`

Product	Supply in tonnes	Demand in tonnes
Petrol	1200	5300
Heating oil	4100	3900
Aviation fuel	8500	2500

Suggest what decision might be taken to match the supply of fuels with the demand.

13 The viscosity of three unknown alkanes was compared: **D**: 1, **E**: 1.2 **F**: 1.6. The higher the number the greater the viscosity. Explain which has the highest boiling point.   `2 Marks`

More challenging

14 Work out the formula for an alkane with 20 carbons. `1 Mark`

15 Describe the colour change that takes place when an alkene is tested with a liquid test reagent. `2 Marks`

16 Describe the trends in viscosity and flammability as the number of carbon atoms in a series of alkanes increases. `2 Marks`

17 Write a balanced equation for the incomplete combustion of heptane, C_7H_{16}. `2 Marks`

18 Explain why there is a need to crack large alkane molecules from crude oil. `2 Marks`

Most demanding

19 An alkane burns in oxygen to produce $7CO_2$. Work out the formula for the alkane and write a full balanced equation for its complete combustion. `4 Marks`

20 Complete the data table for the alkanes below. For each alkane explain its state at room temperature. `4 Marks`

Number of carbon atoms	Formula of alkane	Structure	Boiling point
1			−161
3			−42
5			36
15			270

21 Explain whether heptane, C_7H_{16}, or octane, C_8H_{18}, has the higher boiling point. `2 Marks`

`Total 40 Marks`

CHEMICAL ANALYSIS

IDEAS YOU HAVE MET BEFORE:

PURE SUBSTANCES AND MIXTURES

- Salt and sand can be separated by filtering.
- Iron and sulfur are separated physically by a magnet.
- Crystals can grow when a warm saturated solution cools.

SEPARATING AND TESTING SUBSTANCES

- Chromatography separates inks and coloured dyes.
- Distillation separates alcohol and water.
- Filtering removes solid impurities.

FORMULATIONS OF FOODS

- Orange squash needs water added to it to be drinkable.
- Food products in packets have labels showing the ingredients.
- Amounts of sugar, salt and fats are identified on food labels.

TESTING GASES

- Oxygen relights a glowing splint.
- Hydrogen pops with a burning splint.
- Carbon dioxide turns limewater milky.

USING ANALYTICAL SKILLS

- Chromatography can be used to separate the dyes in sweets.
- Distillation can used to separate alcohol and water.
- Temperature curves can be used to find melting points.

IN THIS CHAPTER YOU WILL FIND OUT ABOUT:

HOW CAN WE TELL IF A SUBSTANCE IS PURE?

- A pure substance has a specific melting point and boiling point.
- Mixtures can be separated by distillation or filtration.
- Pure substances can be obtained by crystallisation.

HOW CAN WE SEPARATE A SUBSTANCE TO ANALYSE IT?

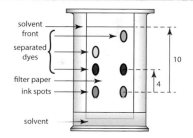

- Dyes travel different distances along chromatography paper.
- Solvent front distances can be measured.
- Ratios of spot position to solvent front can be calculated.

WHAT PRODUCTS NEED TO BE FORMULATED?

- Medicines need to be formulated to exact proportions.
- Fertilisers are formulated according to the job needed.
- Paint and cement must be formulated for consistency.

HOW DO WE TEST FOR GASES FOR ANALYSIS?

- Carbon dioxide turns limewater milky.
- Chlorine bleaches damp litmus paper.
- Hydrogen burns in air with a squeaky pop.

HOW CAN TECHNIQUES BE USED IN ANALYSIS?

- R_f values can be calculated from chromatograms.
- Melting points can be used to establish purity.
- Colours of gases can be used to distinguish the halogens.

KEY CONCEPT

Pure substances

Learning objectives:

- describe, explain and exemplify processes of separation
- suggest separation and purification techniques for mixtures
- distinguish pure and impure substances using melting point and boiling point data.

KEY WORDS

pure
impure
melting point
boiling point

Physical separation

In everyday language, a **pure** substance can mean a substance that has had nothing added to it, so it is in its natural state, for example, pure milk. But this is not the chemical definition.

In chemistry, a *pure substance* is a single element or compound, not mixed with any other substance.

A *mixture* consists of two or more elements or compounds not chemically combined together. The chemical properties of each substance in the mixture are unchanged.

Mixtures can be separated by *physical processes*. These processes *do not* involve chemical reactions.

Separation processes include:

- filtration
- crystallisation
- distillation
- fractional distillation
- chromatography.

1 Explain how you would separate salt and sand.

2 Explain how you would find out the colours in an ink mixture.

Figure 8.1 Separating 'pure' milk. This is not the chemistry definition of pure.

DID YOU KNOW?

Pure and impure milk can both be separated into their components by centrifuging. This means spinning round at high speeds so that layers of substances can be separated. Blood can also be separated in this way.

What can we separate?

We can separate solids from solutions by *filtering*, for example we can filter out excess magnesium metal if we are making a solution of magnesium sulfate from magnesium and sulfuric acid. We can *crystallise* out samples of solids made to assess their purity.

Two liquids with similar boiling points can be separated by *distillation*, for example, ethanol and water. *Fractional distillation* can be used to separate more than two liquids mixed together.

Chromatography can be used to separate dye colours, food colours, inks and different amino acids and to assess the purity of drugs and medicines.

3 **Explain why sulfuric acid is made the limiting reagent when it reacts with magnesium to produce magnesium sulfate.**

4 **A liquid mixture of ethanol and water was found in an old bottle. Insoluble sediment and grit was present. Explain how you would separate the components of the mixture.**

Pure and impure substances

Pure elements and compounds melt and boil at specific temperatures. **Melting point** and **boiling point** data can be used to distinguish pure substances from mixtures.

If a substance is **impure** the melting point will be *lower* and the substance will melt over a *broad* range of temperatures. If a substance is *pure* it will melt more *sharply* at a specific temperature.

If a substance is *impure* the boiling point will be *higher* than the pure substance.

5 **Sam and Alex each made a sample of the same substance. They measured the melting point. They then recrystallised their solid and measured the melting point again. They did this several times.**

| Sam | 46.5 | 46.6 | 46.7 | 46.7 | 46.7 |
| Alex | 42.8 | 43.7 | 44.5 | 44.9 | 45.1 |

Identify who made the purest sample and explain your reasoning.

6 **Akira and Ben tested their samples for their boiling points. The standard boiling point from data tables was 85.0 °C. Their results were: Akira 86.2 °C, Ben 87.1 °C. Identify who had the purest sample and explain your reasoning.**

7 **A beaker of distilled water was thought to have been contaminated with an impurity. The boiling point of the water was 108 °C. The contaminant was separated and found to consist of white crystals. The melting point of the crystals was 190–210 °C.**

   a Explain whether the distilled water was contaminated.

   b Suggest how the contaminant might have been separated from the water.

   c Explain whether the contaminant was pure.

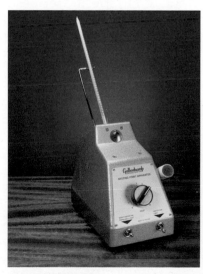

Figure 8.2 Using a melting point to establish purity

Formulations

KEY WORDS
....

formulation
fertilisers
active ingredient
alloys

Learning objectives:

- identify formulations given appropriate information
- explain the particular purpose of each chemical in a mixture
- explain how quantities are carefully measured for formulation.

Medicines have to have an active ingredient but need to have other components mixed with it so that it can easily be taken. This is called a formulation. Why is it necessary to have a medicine formulated and are there other products that need this kind of mixing together?

Products using formulation

A **formulation** is a mixture that has been designed as a useful product. Formulations include food products. They are formulated so that they are consistent and the shopper will know that what they are buying will taste and feel the same every time they buy it. Who would want to buy tomato ketchup that tasted differently each time?

Can you think of other types of products that have to be consistent and are mixtures of specific components?

Paints, **alloys** and **fertilisers** are all substances that are formulated to specific recipes so that they can be used consistently, knowing that the formula used is always the same. Paints have to have the same ingredients, so that the contents of all the tins are the same. It would be useless to paint a room with several pots of paint and find that the colour was different from each pot.

Figure 8.3 This food label shows the ingredients that have formulated this food.

1. Look at the label in Figure 8.3. Identify the ingredients with the most mass and the least mass.

2. Portland cement contains calcium oxide, silica, aluminium oxide, iron oxide and gypsum. Explain why it is very important to mix these in the correct proportions.

Purposes of component chemicals

Iron is a pure metal but it is fairly brittle and rusts easily. By making a formulation with other chemicals it can be made into an alloy. Different formulations make different alloys for different uses. The alloys are usually steel, which is iron mixed with carbon.

Fertilisers are formulated to add nutrients to the soil for different purposes. Some lawn fertilisers may have more nitrogen for shoot growth, some have phosphorus for strong root growth and some have potassium for withstanding drought and disease. Extra magnesium is sometimes in a

KEY INFORMATION
....

The ingredients are listed in order by mass. There is a bigger quantity of the first ingredient.

fertiliser for use if leaves are turning yellow. Magnesium is an essential component of chlorophyll.

There are whole ranges of cleaning agents that are formulated to tackle different kinds of cleaning needs. Some cleaners that tackle grease need to be formulated differently if the grease is on a metal surface or if it is on a granite surface. Cleaners for sinks and bathrooms need a completely different formulation as they need to remove bacteria not grease.

Many products are, therefore, complex mixtures in which each chemical has a particular purpose.

Figure 8.4 Fertilisers are formulated to have different amounts of nutrients such as nitrogen, phosphorus and potassium.

3   Gold is often mixed with silver, copper and other metals. One of its uses is in electrical circuits. Explain why there are a variety of formulations and why they need to be precisely prepared.

4   NPK fertilisers are sold in a formulation by ratio of mass of these three main ingredients. Explain the difference between these two fertiliser formulations.

NPK   4 : 1 : 3   NPK   4 : 2 : 2

Measuring quantities

Fuels, cleaning agents, paints, medicines, alloys, fertilisers and foods are all chemicals that need to be formulated. Formulations are made by mixing the components in carefully measured quantities to ensure that the product has the required properties.

The careful measurement is especially important in the formulation of medicines. The amount of **active ingredient** needed may be very small. This small amount may be difficult to take into the body and difficult to sell. The active ingredient is added to a filler that bulks up the product to a sensible size and a lubricant is added, so that the tablet can then be easily taken. Pills, tablets, capsules and chewable tablets are all made in this way. The filler materials are called excipients. For example, aspirin may contain the excipients starch, lactose and talc.

As each tablet and capsule has to have exactly the same formula there are strict quality assurance processes that are put in place. These processes are laboratory tests that make sure that each tablet is the same and the quality of the ingredients is at the correct standard.

DID YOU KNOW?

In some countries there are different formulations of winter and summer petrol. They have different ratios of the ingredients so that the performance of the petrol is maximised for the particular season. Some volatile hydrocarbons are removed in summer.

REMEMBER!

You do not need to know the names of components in proprietary products.

5   Some medicines are quality assured for correct formulation.

Medicine	A	B	C	D	E	F
% active ingredient	6.25	6.25	6.25	6.25	6.27	6.25
% filler	69.27	69.27	69.37	69.27	69.25	69.27
% lubricant	24.48	24.48	24.38	24.48	24.48	24.48

Explain which medicines would be rejected.

6   Most paints used to be formulated with a solvent and oil base. In the last two decades water-based paints were developed using new techniques. Suggest why the paints were developed with this different formulation.

Chromatography

KEY WORDS

chromatography
R_f value
stationary phase
mobile phase

Learning objectives:

- explain how to set up paper chromatography
- distinguish pure from impure substances
- interpret chromatograms and determine R_f values.

You will have already seen how chromatography with paper can be used to separate coloured dyes. It can also be used to identify colourless mixtures and to extract small amounts of pure substances such as medicines from plants. How can we use it to identify substances or solve mysteries?

Separating dyes

You will already have put spots of dye on **chromatography** paper and fixed the edge in water to watch the colour rise up through it.

This was a simple example where the chromatography paper was the **stationary phase** and the water was the **mobile phase**.

An example of how this technique can be used to solve questions is in testing food colourings. A spot of each standard food colour is dropped onto the start line. A spot of the unknown food colour mixture is dropped on the line at the end of the paper. The sheet is held in the solvent (water) and left for separation to take place. The resulting chromatogram could be as in Figure 8.6.

You can see that the four standard colours each have one spot that has travelled up the paper. This distance can be measured. The food colour mixture has three colours:

- the yellow spot has moved the same distance as the E102 yellow spot
- the light blue spot is the same distance as the E133 light blue spot
- there is also a green spot.

Has the green spot moved the same distance as the E142 green spot? The answer is no, so E142 is not in the mixture, it is another substance.

1 **Explain how you know there is no E131 in the test mixture.**

2 **Explain how you would find out which green colour is in the food.**

Measuring R_f values

In the example above it was clear to see that the colours had moved different distances as they were all on the same chromatogram. A standard measure needs to be used so that

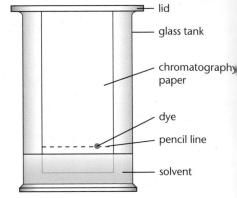
Figure 8.5 Carrying out chromatography

lid
glass tank
chromatography paper
dye
pencil line
solvent

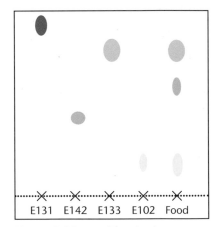
E131 E142 E133 E102 Food

Figure 8.6 Dyes of food colours.

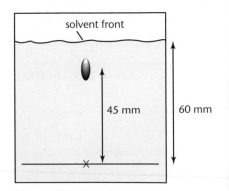
solvent front
45 mm
60 mm

Figure 8.7 Measuring R_f values

separation can be compared in any situation. This is the R_f **value** where:

$$R_f = \frac{\text{distance moved by substance}}{\text{distance moved by solvent}}$$

This is the ratio of the distance moved by a compound (from the origin to the centre of the spot) to the distance moved by the solvent.

In the example in Figure 8.7, the substance has moved 45 mm and the solvent has moved 60 mm.

$$R_f = \frac{45}{60} = 0.75$$

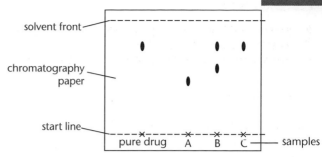

Figure 8.8 Chromatography to identify drug sample

Different compounds have different R_f values in different solvents, which can be used to help identify the compounds.

In paper chromatography a solvent moves through the paper carrying different compounds different distances, depending on their attraction for the paper or the solvent.

3 Calculate the R_f value for a dye that moves 56 mm in a solvent that moves 70 mm.

4 The R_f value for a blue dye is 0.68 and the solvent front moves 90 mm. How far should a blue dye spot move?

Assessing purity of drugs

In chromatography, separation depends on the distribution of substances between the stationary and mobile phases. Impure substances or mixtures will separate into their component parts. This happens with paper chromatography or with thin layer chromatography, where a white solid thinly spread on a glass slide is used.

Thin layer chromatography is a method for finding the purity of a compound. The R_f value of a sample is compared against the R_f of a sample of substance known to be pure.

The compounds in a mixture may separate into different spots depending on the solvent but a pure compound will produce a single spot in all solvents. Because of this, chromatography can be used to assess the purity of compounds such as drugs and medicines. For example, a chromatogram such as Figure 8.8 can be developed.

Use Figure 8.8 to answer these questions:

5 Explain which sample (A, B or C) is the purest drug.

6 Calculate the R_f value of each of the spots of B on the chromatogram.

KEY SKILL

In Figure 8.8 the solvent used is water. Remember that the edge of the paper needs to just dip in the water. The spot must not be in the water.

DID YOU KNOW?

The pure substance moves along the white solid of the stationary phase (usually silica gel or aluminium oxide) and can be scraped off at the end. It can then be extracted from the solid to be used as a pure substance. This is how we can extract small amounts of useful medicines from rare plants.

REQUIRED PRACTICAL

Investigate how paper chromatography can be used in forensic science to identify an ink mixture used in a forgery

KEY WORDS

chromatogram
solvent front
chromatography

Learning objectives:

- describe the safe and correct manipulation of chromatography apparatus and how accurate measurements are achieved
- make and record measurements used in paper chromatography
- calculate R_f values.

Interpreting evidence is an important skill for forensic scientists. Chromatography can be used to separate mixtures, such as ink mixtures, so that patterns of separate inks appear as 'spots' on a chromatogram. Each separate ink travels a different distance when in the same solvent. Could you identify which colours make up the mixed ink dye used in a forgery?

A number of different skills are needed to carry out an identification using **chromatography**, including manual dexterity and measuring. This topic looks at the skills needed to identify inks in a mixture.

❗ These pages are designed to help you think about aspects of the investigation rather than to guide you through it step by step.

DID YOU KNOW?

Paper chromatography is only one chromatographic technique that can be used: there is also thin layer chromatography and gas-liquid chromatography.

Safe and correct use of apparatus

Figure 8.9 shows a paper **chromatography** experiment. Ink spots are placed along a start line and held in a small volume of water which acts as the solvent.

Think about these questions:

1 The start line is drawn 1 cm from the bottom of the paper. Explain why the start line is drawn in pencil and not ink.

2 Name the piece of apparatus used to make the ink spots on the paper.

3 Explain why the level of the solvent should be below the ink spots.

4 Suggest when the developing chromatogram should be removed from the solvent.

Once the solvent has risen up through the chromatography paper the **chromatogram** is dried before measurement.

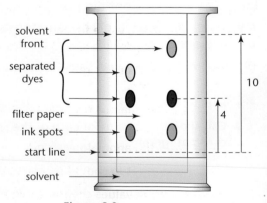

Figure 8.9

Sam and Alex were given an ink mixture that was used in a forgery and asked to identify which separate inks had made up the mixture.

They put spots of the separate ink along the start line, added a spot of ink mixture at the end of the paper and developed the chromatogram. They used water as the solvent to 'elute' the spots.

Making and recording results

From the start line Sam and Alex needed to measure:

- the distance the solvent had travelled (the **solvent front**)
- the distance each spot of ink had travelled.

They recorded these distances from the start line:

Chromatograms	Distance of red spot cm	Distance of blue spot cm	Distance of yellow spot cm	Distance of green spot cm	Distance of solvent cm	Distance of mixture spots cm
Sam's	4.2	5.4	6.8	10.3	12.4	4.3, 5.3, 10.4
Alex's	4.1	5.6	6.7	10.1	12.5	4.4, 5.5, 9.9
Jo's		4.5	7.8	8.1	10.0	3.2, 4.4, 8.2

5 Identify the colours that made up the forgery ink mixture.

6 Name the colour that was not part of the mixture.

Calculating R_f values

As the solvent front may not move the same distance each time, it important to calculate the ratio of distances travelled by solvent and ink. This is given as an index known as the R_f value.

$$R_f = \frac{\text{distance travelled by ink spot}}{\text{distance travelled by solvent front}}$$

7 Calculate the R_f value for the yellow spot on Alex's chromatogram.

Jo also made a chromatogram with the ink mixture and separate inks but did not leave the chromatogram in for as long. The solvent front did not travel as far as Sam's or Alex's.

8 Show that the R_f value for Jo's blue spot is consistent with the R_f value for Alex's blue spot.

9 Predict a value for Jo's red spot using an R_f value calculated from Sam's data.

10 Jo has one anomalous result (a result that does not fit the pattern).

  a Identify the anomalous result.

  b Explain what Jo should do about this result.

11 Suggest why R_f values change when the solvent is changed.

KEY INFORMATION

To get a *reproducible* result the data should be able to be obtained by another experimenter. Sam and Alex have similar results. Jo has similar results too, except for one anomalous result that does not fit the pattern started by Sam and Alex.

Testing for gases

Learning objectives:

- recall the tests for four common gases
- identify the four common gases using these tests
- explain why limewater can be used for testing CO_2.

KEY WORDS

carbon dioxide
chlorine
hydrogen
oxygen

Gases are released in many chemical reactions. The gases we will meet regularly are hydrogen, oxygen, carbon dioxide and chlorine. Each one can be tested for in a specific way. Have you tried testing for any of these gases yet?

Testing for hydrogen and oxygen

The test for **hydrogen** uses a burning splint held at the open end of a test tube of the gas. Hydrogen burns rapidly with a 'pop' sound.

The basis for this test is that hydrogen is reacting rapidly as a fuel by burning in oxygen. The product is water.

The test for **oxygen** uses a glowing splint inserted into a test tube of the gas. The glowing splint relights in oxygen.

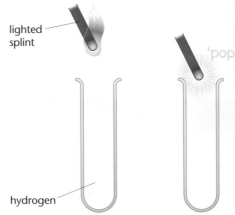

Figure 8.10 Testing for hydrogen

Figure 8.11 Testing for oxygen

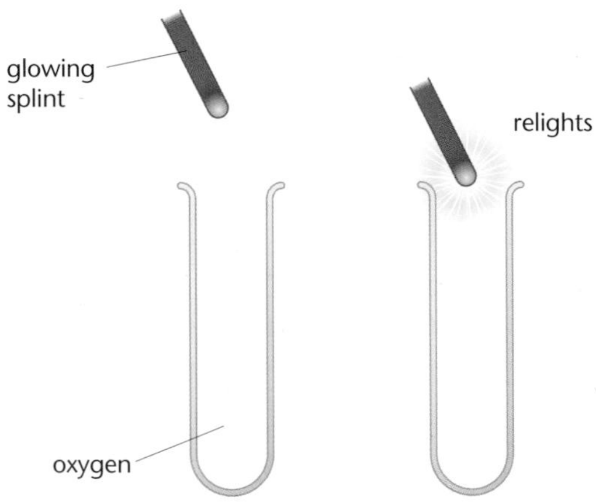

1. **Write a word equation for the reaction when hydrogen goes 'pop'.**
2. **Why can oxygen relight a glowing splint? (Hint: think about the combustion triangle. What are the three conditions that are needed for a flame?)**

Testing for carbon dioxide

The test for **carbon dioxide** uses an aqueous solution of calcium hydroxide (which is often called 'limewater'). When carbon dioxide is shaken with or bubbled through limewater, the limewater turns milky (it looks cloudy).

Figure 8.12 Testing for carbon dioxide

DID YOU KNOW?

Some airships used to be filled with hydrogen. The Hindenburg airship caught fire in May 1937, with disastrous consequences, as it tried to moor in New York after a voyage across the Atlantic.

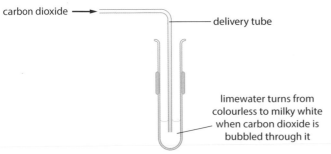

What is happening with this test? Limewater is calcium hydroxide. When carbon dioxide is passed through limewater, the product formed is calcium carbonate. This white chalky substance gives a milky appearance to the solution.

③ Write the word equation for the reaction between limewater and carbon dioxide.

④ Find out why the limewater turns colourless again if CO_2 continues to be passed through the limewater.

Testing for chlorine

The test for **chlorine** uses litmus paper. When damp litmus paper is put into chlorine gas the litmus paper is bleached and turns white. Chlorine can also be bubbled into water and tested with litmus paper, which bleaches.

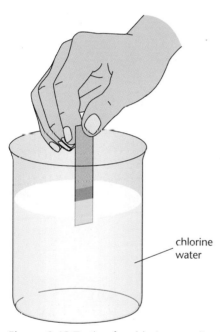

chlorine water

Figure 8.13 Testing for chlorine gas dissolved in water

⑤ What colour is chlorine gas?

⑥ Which halogen gives a brown-red gas when warmed?

⑦ Which halogen gives a purple gas when warmed?

Use an appropriate number of significant figures

Learning objectives:

- measure distances on chromatograms
- calculate R_f values
- record R_f values to an appropriate number of significant figures.

Chromatography can be used to separate mixtures using different solvents. As the solvent moves along a piece of chromatography paper, different parts of the mixture travel different distances with it.

Can you remember how to calculate the R_f value of a spot on a chromatogram?

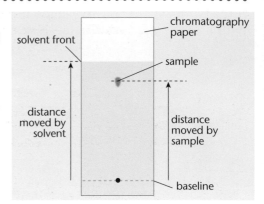

Figure 8.14

Measuring distances

There are two distances that need measuring on a developed **chromatogram**:

- the distance from the baseline to the solvent front
- the distance from the baseline to the spot from the mixture.

Both these can be measured with a ruler in cm or mm. Is your technique or your ruler accurate enough to measure these distances to 1/10th of a millimetre (0.1 mm)? Probably not, although you may be able to see to 0.5 mm. But is the apparatus set up well enough to achieve this level of accuracy? Probably not.

Look at Figure 8.14 and use it to answer these questions.

1 What is the distance from the baseline to the solvent front in mm?

2 What is the distance from the baseline to the middle of the sample spot in mm?

Calculating R_f values

To make sure that we can identify a particular chemical in a solvent by using a chromatogram we need to calculate the **ratio** of the two distances.

The reason we do this is because whatever distance the spot travels, it will always be in the same **proportion** to the distance travelled by the same solvent.

$$R_f = \frac{\text{distance travelled by the spot}}{\text{distance travelled by the solvent}}$$

DID YOU KNOW?

Accuracy and precision have different meanings in science.

Precise measurements are those where there is very little spread from the mean.

Accurate measurements mean that your measurements are very close to a standard or accepted value.

For example, if the spot travels 8 cm and the solvent travels 10 cm then the ratio is 0.8.

This ratio will be the same if we measure in mm. Now the distances are 80 mm and 100 mm and again the ratio is 0.8.

If you had been short of time in your practical lesson and only allowed the solvent to travel 5 cm, then we would notice that the spot of the same sample chemical would travel 4 cm. Again the ratio would be 4/5 or 0.8.

Look back again at the chromatogram in Figure 8.14. Can you calculate the **Rf value**?

Now look at Figure 8.15 and use it to answer these questions.

3 Measure the distance of the solvent front from the baseline. Also measure the distance travelled by each of the five spots.

4 Calculate the R_f value of each of the spots and suggest, with reasons, which of the samples (A, B or C) is likely to be the pure drug.

Applying significant figures

When we have calculated the R_f values, to how many significant figures should we record the result?

Here are examples of values measured on three chromatograms:

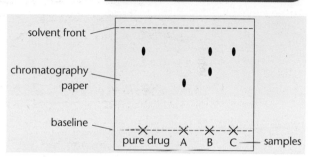

Figure 8.15 A paper chromatogram for measuring drug purity

Experiment number	Distance travelled by sample spot (cm)	Distance travelled by solvent (cm)	R_f value shown on calculator	R_f value to sig fig
1	4	8	0.5	0.5
2	13.5	18.0	0.75	0.75
3	4.1	8.6	0.47674418	0.48

- In experiment 1, each value is measured to 1 significant figure (sig fig), so the R_f is calculated to 1 sig fig.
- In experiment 2, one measurement value is given to 3 sig figs while the other has 2 sig figs. The R_f value is written with 2 sig figs, which is the same as the value with the *lowest* number of sig figs.
- In experiment 3, both measurement values have 2 sig figs so the R_f value is written with 2 sig figs. Note that the result needs to be rounded up to 0.48.
- Now use the values you calculated in Question 4 (from Figure 8.15) to answer the next questions.

The third place value (6) in 0.476 is greater than 5 so the second place value (7) is rounded up to 8.

5 What is the R_f value of sample C, to the appropriate number of significant figures? Explain why you chose this number of significant figures.

6 Record the R_f values for samples A and B to the appropriate number of significant figures.

Check your progress

You should be able to:

describe, explain and identify examples of processes of separation such as filtration, crystallisation and distillation → suggest separation and purification techniques for mixtures → distinguish pure and impure substances using melting point and boiling point data

describe how to set up paper chromatography → distinguish pure from impure substances → interpret chromatograms and determine R_f values

describe the tests for oxygen and hydrogen → describe the test for chlorine → describe the test for carbon dioxide

Worked example

1 **Suggest which piece of apparatus you would use to separate sand from alcohol.**

Filter funnel and filter paper

The answer is correct.

2 **Suggest the difference between the melting point of a pure and an impure substance.**

An impure substance has a lower melting point than a pure one.

The answer is correct

3 **Draw the spots that will appear on the chromatogram if the mixed dye contains colours Q, R and T.**

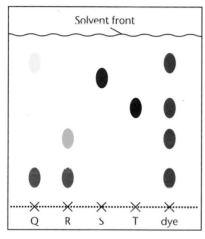

Solvent front

Q   R   S   T   dye

This answer is correct as the distances of all the spots are the same as the originals, and the red spot is not there.

4 **Calculate the R_f value for the red spot S.**

R_f value = solvent distance/spot distance
= 14/10.5 = 1.3 mm

This answer is incorrect as the R_f value is spot distance /solvent distance. Also there are no units to an R_f value. The answer should have been R_f = 10/14.5 = 0.69

5 **These reactions give off gases. Complete the table.**

Reaction	Heating carbonates	Acids on metals	Catalyst on hydrogen peroxide	Electrolysis of concentrated NaCl
Gas	CO_2	hydrogen		Cl_2
Test and result	Turns limewater milky	Pops lighted splint	Relights glowing splint	Turns litmus red

The answers in the first two columns are correct. There is an omission in the third column where the student forgot to write oxygen or O_2. The last response is incorrect, it should be damp litmus paper used which will be *bleached*, not turns red.

End of chapter questions

Getting started

1 Describe what you would see if you held a glowing splint in a tube of oxygen. `1 Mark`

2 Describe what you would see if you held a lighted splint in a tube of hydrogen. `1 Mark`

3 Fertilisers need to be made up to the correct 'recipe'. Write down two other products that need to be formulated. `1 Mark`

4 Alex wanted to finish painting a room in the same colour. What would be most important about the next tin of paint used?

a the same size   b the same formulation   c the same volume   d the same mass `1 Mark`

5 A tablet is made of 0.050 g of aspirin added to 0.79 g of starch and 0.11 g of talc. Calculate the percentage of active ingredient in the tablet. `1 Mark`

6 Sand has been added to a solution of black ink. Describe how you would separate the sand from the solution and find which colours were in the ink. `2 Marks`

7 Three food packet labels A, B and C contain this information on their formulation. `1 Mark`

	Mass per 100g food A	Mass per 100g food B	Mass per 100g food C
Starch	40		50
Sugar	20		25
Saturated fat	20		16
Unsaturated fat	9		8.5
Flavouring	1.0	0.8	0.5

Food B had:

- 7 more g of starch than food A.

- 3 less g of sugar than food C.

- Equal amounts of saturated and unsaturated fat.

Fill in the data for food B.

Going further

8 Explain the method for separating ethanol (Bp 80 °C) and water (Bp 100 °C).

1 Mark

9 Aspirin is a formulation. Explain why the substances in aspirin have to be mixed in exactly the correct proportions.

1 Mark

10 Describe how to test for chlorine gas.

2 Marks

11 Some pure citric acid crystals are needed from a mixture of citric acid, sand and alcohol. Explain how the sand is separated from the mixture, then the alcohol, and finally how the citric acid is obtained and assessed for purity.

4 Marks

12 The Contact process is used to make sulfuric acid. Sulfur trioxide, SO_3, is made as part of the process.

$$2SO_2 + O_2 \rightleftharpoons 2SO_3$$

  a Work out the molar masses of the reactants and product.

  b Calculate the number of moles in 800 g of SO_3.

4 Marks

13 Cleaning agent powders are formulated with different amounts of ingredients. The data gives the mass of each ingredient in 100 g of cleaning agents W, X, Y and Z.

Cleaning agent	W	X	Y	Z
Surfactant/cleaner	94.6	93.3	84.4	92.8
Optical brightener	0.7	0.4	0.1	0.5
Foam regulator	1.1	1.7	5.2	1.9
Fragrance	0.2	0.4	0.9	0.1
Fabric softener	2.6	2.4	5.8	2.5
Water softener	0.8	1.8	3.6	1.5

  a Which powder is formulated very differently from the other three?

  b Which ingredients does it have less of than the other powders?

  c Which powder will make material cleaned seem brighter?

  d Which powder will smell the least?

2 Marks

More challenging

14 Which technique is used to separate a solid from a solution? `1 Mark`

15 Which technique is used to separate two liquids with different boiling points? `1 Mark`

16 Identify the stationary phase and the mobile phase in paper chromatography. `1 Mark`

17 Describe two observations seen when passing carbon dioxide through limewater. `2 Marks`

18 Identify four of the following substances that require formulation: `4 Marks`

Fertiliser   Diamond   Crude oil   Cement

Aluminium Paint   Alloy   Sulfur

Most demanding

19 Define an R_f value

1 Mark

20 Use Figure 8.6 to calculate the R_f values for:

4 Marks

   a E131   b E142   c the green spot in the food.

Explain why the green spot in the food is not E142.

21 Gloss paint, matt paint and water-based paint have different formulations. Gloss paint does not contain extenders. Water-based paint has less pigment. Evaluate the data for X, Y and Z and identify which type of paints they are.

4 Marks

Paint	% solvent	% binder	% additives	% pigment	% extender
X	40	30	5	25	0
Y	44	14	5	25	12
Z	60	20	7	13	0

Total: 40 Marks

THE ATMOSPHERE

VOLCANOES ERUPT PRODUCING GASES

- Volcanoes form on gaps between tectonic plates.
- Volcanoes produce lava and many gases.
- The Earth was very hot when it was formed, and it is cooling.

PLANTS USE CARBON DIOXIDE IN PHOTOSYNTHESIS

- Plants photosynthesise using carbon dioxide and water.
- Plants make oxygen, which is used by animals to keep them alive.
- Plants need sunlight for photosynthesis.

GLOBAL WARMING MELTS ICE CAPS

- There are different climates in different parts of the world.
- The Earth's surface is constantly changing.
- Animals have difficulty surviving with habitat loss.

USING LESS FOSSIL FUEL CUTS DOWN POLLUTION

- Cars, ships and planes use fossil fuels to transport us.
- We use fossil fuels to heat buildings and drive machinery.
- Burning fossil fuels causes air pollution, which is bad for health.

ATMOSPHERIC POLLUTION

- Exhaust fumes can cause breathing problems.
- Acid rain can cause damage to buildings and trees.
- Soot damaged buildings are being cleaned.

9

IN THIS CHAPTER YOU WILL FIND OUT ABOUT:

WHAT WAS THE EARTH'S EARLY ATMOSPHERE LIKE?

- The early atmosphere arose from gases from volcanoes.
- Water vapour condensed to form the oceans.
- There was a high percentage of carbon dioxide and no oxygen.

WHY DID THE EARLY ATMOSPHERE CHANGE?

- Algae used carbon dioxide for photosynthesis.
- Photosynthesis produced oxygen so levels increased.
- Oceans dissolved carbon dioxide, making them acidic.

WHAT ARE THE CONSEQUENCES OF THE GREENHOUSE EFFECT?

- Greenhouse gases are essential for keeping temperatures stable.
- If we had no greenhouse effect the Earth would not sustain life.
- The balance of gases changes the effect and could be damaging.

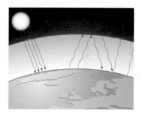

HOW CAN WE REDUCE THE EFFECT OF HUMAN ACTIVITY?

- We can reduce the use of fossil fuels and use renewable energy.
- We can use sunlight more directly for energy needs.
- We can use resources more efficiently and fairly.

WHAT ARE THE EFFECTS OF ATMOSPHERIC POLLUTION?

- Sulfur dioxide dissolves in water to form acid rain.
- Sulfur dioxide and oxides of nitrogen can cause respiratory problems.
- Particulates can cause global dimming.

Proportions of gases in the atmosphere

Learning objectives:

- identify the gases of the atmosphere
- recall the proportions of gases
- explain how the balance of the gases is maintained.

KEY WORDS

atmosphere
carbon dioxide
nitrogen
oxygen

The atmosphere was not always made up of the gases in the mixture that we know today. However, for the last 200 million years, the proportions of different gases in the atmosphere have been much the same as they are today. Without oxygen and carbon dioxide, life as we know it would not be the same.

Air as a mixture

Air is a mixture of different gases. Clean air contains **nitrogen**, **oxygen**, **carbon dioxide**, water vapour and noble gases, such as argon.

The amount of water vapour changes but the amounts of the other gases remain almost constant.

The amounts of the various gases in clean air remain constant because of two opposing processes:

KEY INFORMATION

The amount of water in the **atmosphere** in different parts of the world gives rise to dry desert conditions or to warm, wet and humid climates.

Photosynthesis	Respiration
plants use carbon dioxide and release oxygen	animals and plants use oxygen and release carbon dioxide

These two processes help to keep the balance of the proportions of gases in the air.

1. Describe how oxygen is released into the atmosphere.

2. Suggest the proportion of water vapour in the atmosphere for the following conditions:

   a dry desert   **b** hot rain forest   **c** cool maritime.

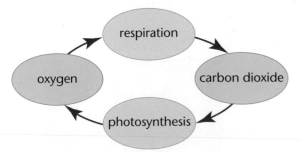

Figure 9.1 Plants photosynthesise, releasing oxygen. Animals and plants respire, releasing carbon dioxide.

Proportions

In the mixture of gases in the air there is:

- approximately 80% nitrogen (about four-fifths or $\frac{4}{5}$)

- approximately 20% oxygen (about one-fifth or $\frac{1}{5}$)

- a small proportion of carbon dioxide (approximately 0.040%, often given as 400 parts per million)
- variable amounts of water vapour
- small proportions of noble gases.

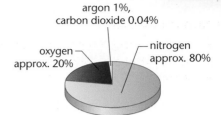

Figure 9.2 Proportions of gases in dry air

The processes of photosynthesis by plants and respiration by plants and animals help to keep the balance of the proportions of oxygen and carbon dioxide in the air. Nitrogen is a very unreactive gas so its proportion in the air is kept constant.

3 **Explain why the atmosphere of the planet Mars has only traces of oxygen.**

Balancing the proportions

The percentages of gases in the air do not change very much. This is because there is a balance between processes that use up carbon dioxide and make oxygen and those that use up oxygen and make carbon dioxide.

Plants use carbon dioxide to make glucose through photosynthesis:

carbon dioxide + water → glucose + oxygen

$6CO_2$ + $6H_2O$ → $C_6H_{12}O_6$ + $6O_2$

Animals and plants use glucose and oxygen through respiration to release energy:

glucose + oxygen → carbon dioxide + water

$C_6H_{12}O_6$ + $6O_2$ → $6CO_2$ + $6H_2O$

These two processes are part of the carbon cycle. One other process that is part of this cycle is also part of the balance of the gases. This process is combustion. Combustion uses oxygen and produces carbon dioxide.

fuel + oxygen → carbon dioxide + water

4 **Breathing and respiration are different processes. Explain the difference, including the gases involved.**

5 **Explain why 6 molecules of CO_2 are produced from 1 molecule of glucose.**

6 **Explain the role of glucose in maintaining the Earth's atmosphere.**

DID YOU KNOW?

The level of carbon dioxide is constantly monitored at Mauna Loa, a volcano in Hawaii. Measurements are usually given in ppm (parts per million). It is monitored because the rising levels of carbon dioxide are causing great concern. Scientists have been measuring concentrations of CO_2 at Mauna Loa since 1958.

The Earth's early atmosphere

Learning objectives:

- describe ideas about the Earth's early atmosphere
- interpret evidence about the Earth's early atmosphere
- evaluate different theories about the Earth's early atmosphere.

KEY WORDS

atmosphere
oceans
sediments
volcanic

The age of the Earth is currently estimated to be between 4.5 and 4.6 billion years. That is a little over four and half thousand million years (often written as 4.5×10^9 years). During that time the atmosphere has changed from possibly hydrogen and helium at first, probably followed by gases from volcanic activity, to finally an atmosphere rich in oxygen, which is what we have now.

The Earth's early atmosphere

We *definitely know* we now have an **atmosphere** rich in oxygen but we also *think* we know, from evidence, that oxygen was not in the Earth's early atmosphere. From the evidence we have, scientists have developed theories about the change in atmosphere over the last 4.6 billion years.

One theory suggests that during the first billion years of the Earth's existence there was intense **volcanic** activity that released gases. This process is called degassing and the idea is based on the composition of gases vented out in present-day volcanic activity.

At the start of this period the Earth's atmosphere may have been like the atmospheres of Mars and Venus today, consisting of mainly carbon dioxide with little or no oxygen gas.

Volcanoes also produced nitrogen, which gradually built up in the atmosphere, and there may have been small proportions of methane and ammonia.

These gases formed the early atmosphere, along with water vapour, which condensed to form the oceans.

When the **oceans** formed, carbon dioxide dissolved in the water and carbonates were precipitated producing **sediments**, reducing the amount of carbon dioxide in the atmosphere.

1. Explain the scientists' theory of how the first gases became part of the atmosphere for the first billion years.

2. Suggest why there was no oxygen in the early atmosphere.

3. There are several different theories about the Earth's early atmosphere. Suggest why.

REMEMBER!

You only have to remember this one theory, but there are more theories you can look up and evaluate against evidence.

Figure 9.3 Volcanic activity releases gases.

Figure 9.4 The surface of Mars is being explored by robots but the atmosphere cannot support human life.

Evidence for the theories

The theories about what was in the Earth's early atmosphere and how the atmosphere was formed have changed and developed over time.

One set of evidence is gained by measuring carbon and boron isotope ratios in sediments under the sea.

Other models use the composition of gases given out by volcanoes today as evidence. An assumption is made that the same proportions of gases are given out today as were given out by volcanoes billions of years ago.

The question we need to ask is how well these assumptions fit with evidence gained through other means. We need to *evaluate evidence* to develop new models.

4 Scientists use the composition of gases given off by volcanoes to develop models of the Earth's ancient atmosphere. Suggest one reason why.

Figure 9.5 Water vapour was produced, which condensed to form oceans.

Evaluating the theories

The evidence for the early atmosphere is limited because of the time scale of 4.6 billion years, which means no direct measurements can be made. Models need to be developed using proxy evidence or by assuming that what happened in the past is still happening today and using that evidence from today (for example the composition of volcanic gases).

Proxy evidence is evidence from one source gathered from ancient times that can be used to make an assumption about another related ancient effect. An example would be counting the number of stomata on fossils of ancient leaves to make assumptions about the levels of carbon dioxide in the atmosphere.

Evidence from direct measurements would seem more valuable but there is not always agreement that the measurements were taken correctly. For example, the measurements taken from isotopes by one team are seen as more reliable if their data compares well with the data from a different site by a different team. The more measurements taken by different teams that provide data that coincide, the more reliable the data. This is why scientists publish data in scientific journals.

5 Scientists have only indirect evidence for the atmosphere of 2 billion years ago. Suggest, with reasons, whether the current composition of volcanic gases gives better or worse evidence than the evidence from counting stomata.

6 Stomata are openings that allow gas exchange in plants. Predict a relationship between the number of stomata and atmospheric carbon dioxide levels.

DID YOU KNOW?

Some scientists are now rejecting the early models of the Earth's atmosphere and are studying zircon gemstones for new evidence.

How oxygen increased

Learning objectives:

- identify the processes allowing oxygen levels to increase
- explain the role of algae in the composition of the atmosphere
- recall the equation for photosynthesis.

KEY WORDS

algae
evolve
oxygen
photosynthesis

The ages of the Earth are divided into eons and eras spanning billions of years. The 'Hadean' is the first eon (4.5 to 4 billion years ago). The second eon is the Archaen. During this eon we find the first evidence of organisms that photosynthesise. This material is kerogen, which is the compressed remains of planktons found in shales. When heated, kerogen releases bitumen and crude oil.

The early production of oxygen

We have already seen that the Earth is 4.5 billion years old. In the first billion years the Earth's atmosphere may have been like the atmospheres of Mars and Venus today, consisting of mainly carbon dioxide with little or no **oxygen** gas.

The first life forms appeared over 3.5 billion years ago, but these could live anaerobically, that is without oxygen. **Algae** first produced oxygen about 2.7 billion years ago.

Soon after the appearance of life forms that used **photosynthesis**, oxygen appeared in the atmosphere. Over the next billion years, plants **evolved** and the percentage of oxygen gradually increased to a level that enabled animals to evolve.

Figure 9.6 Algae started to produce oxygen about 2.7 billion years ago.

1 When did algae first produce oxygen?

2 Explain the difference between living anaerobically and living aerobically.

Algae and photosynthetic production of oxygen

Algae and plants produced the oxygen that is now in the atmosphere. This was produced by photosynthesis.

This process can be represented by the equation:

$$6CO_2 + 6H_2O \rightarrow C_6H_{12}O_6 + 6O_2$$

carbon dioxide + water → glucose + oxygen

You can see that the process uses up carbon dioxide. So, as more plants evolved the levels of carbon dioxide went down and the levels of oxygen went up.

Over billions of years, the percentage of oxygen in the atmosphere increased and the percentage of carbon dioxide decreased, until today's levels were reached.

Figure 9.7 These fossils are of early plants that produced oxygen by photosynthesis.

KEY INFORMATION

You can show by experiment that aquatic plants produce oxygen.

The percentage of nitrogen also slowly increased, but since nitrogen is very unreactive, very little nitrogen was removed from the atmosphere.

3. Oxygen is a product of photosynthesis. What is the other product?

4. Suggest why the levels of oxygen increased.

5. Explain the role of algae in the increased levels of oxygen.

The Great Oxygenation event

There is a theory that the first organisms carrying out photosynthesis, cyanobacteria, produced oxygen as a waste product, but that this oxygen was removed by the oxidation of iron to form iron(III) oxide. The evidence for this is banded iron oxide sediments, which are red and nearly always found in older rocks formed by precipitation in oceans. Examples of these formations can be found in Minnesota and Western Australia.

The process of iron oxidation captured all the spare oxygen until there was an excess that could no longer be captured. This led to a significant rise in the levels of oxygen in the atmosphere and is known as the Great Oxygenation event. This is a theory based on evidence, which can be interpreted in different ways and combined with other evidence to support the theory. The evidence cannot be gathered by direct measurement.

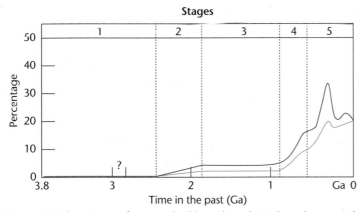

Figure 9.9 The stages of oxygen build-up thought to have happened over billions of years. The unit on the *x* axis (Ga) means 'billions of years'. The two lines show the range of estimates.

One piece of evidence scientists used to develop theories about the rise of oxygen levels still carries on today. This is the presence of microbial mats. They are present in the fossil record of 3500 million years ago but are still present and can produce a thin layer of oxygenated water under thick ice. These microbes do not die in an oxygenated environment or during ice ages.

6. Suggest why there was a time lag between the appearance of the first organisms producing oxygen and the rise in the level of free oxygen in the atmosphere.

Figure 9.8 A banded iron formation in Western Australia which originally used up oxygen from the air before the levels rose.

DID YOU KNOW?

This theory has many different names and is an attempt to explain why there was a delay between the appearance of oxygen-producing bacteria and increases in oxygen levels.

How carbon dioxide decreased

Learning objectives:

- describe the main changes in the atmosphere over time
- describe some of the likely causes of these changes
- explain how the deposits of limestone, coal, crude oil and gas were formed.

KEY WORDS

compressed
fossil fuels
marine life
sedimentary

Carbon dioxide formed a higher percentage of the Earth's early atmosphere than it does now. Over billions of years the level of carbon dioxide decreased. Why did this happen and how do we know it happened? Models of the ancient atmosphere are still being debated.

Trapping carbon dioxide

Algae and plants decreased the percentage of carbon dioxide in the atmosphere by using the process of photosynthesis.

Figure 9.10 Plant remains were compressed, forming coal.

They used carbon dioxide to make glucose, and they also released oxygen.

As the plants and algae grew and proliferated over time the amount of carbon dioxide decreased and the amount of oxygen increased.

As the plants died and decayed, they formed a thick layer of plant deposits, which was **compressed** first to peat and then eventually formed coal.

Coal is formed from the thick plant deposits that were buried and compressed over millions of years. Coal is mostly carbon that became locked in the ground. This resulted in lower levels of carbon dioxide.

Another organism that used up carbon dioxide was plankton. The remains of plankton were deposited in muds on the sea floor and were covered over and compressed over millions of years. This produced crude oil and natural gas, which became trapped in the rocks.

Figure 9.11 Peat and bituminous coal

Both these **fossil fuels** trapped carbon. Coal traps carbon as a **sedimentary** rock and crude oil/natural gas traps carbon between layers of other rock.

1. Describe how coal was formed.

2. Describe two ways that carbon dioxide levels were decreased over billions of years by the formation of fossil fuels.

Forming limestone sediments

Carbon dioxide levels were also further lowered after the appearance of animals, when their shells and skeletons made of calcium carbonate compressed together over time, forming limestone.

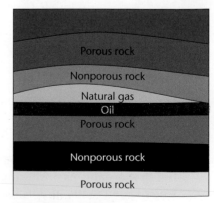

Figure 9.12 Forming oil from plankton compressed in mud over millions of years

Limestone is a sedimentary rock, mainly made of calcium carbonate, which again locked up carbon and added to the lowering of carbon dioxide levels.

Carbon dioxide also dissolves in the oceans to make a very weakly acidic solution and so the levels were, again, lowered by this process.

3 Describe how limestone was formed.

4 Explain how plants were essential for early animal life to flourish.

Evidence of carbon dioxide levels

There is a direct method for measuring the concentration of carbon dioxide in the ancient atmosphere. This is to drill down and take samples of ice cores from Greenland or Antarctica. Bubbles of air are trapped in the ice and their composition can be measured.

Evidence is direct but these levels can only be plotted as far back as 800 000 years.

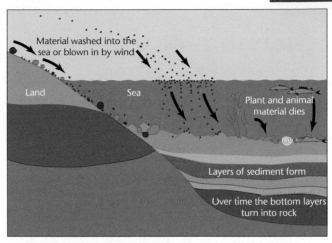

Figure 9.13 Sedimentary rocks are formed as layers are compressed by other layers deposited above.

DID YOU KNOW?

Gas concentrations are measured in parts per million (ppm). Over the last 400 000 years the concentration of CO_2 has followed cycles of a concentration of approximately 180 ppm to 280 ppm.

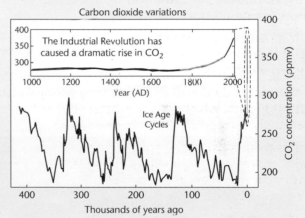

Figure 9.14 Cycles of carbon dioxide levels over the last 400 000 years to 1700 AD

This seems a very long time span but is only a tiny fraction of the time of development of the atmosphere, so how is evidence gathered about the levels further back in time towards 4.5 billion years ago?

This is again done by using proxy evidence. This evidence includes methods such as isotope ratios in **marine life** sediments or again by counting stomata.

Using proxy measurements means that scientists have different ways of looking at the evidence and there is great debate about the interpretation of what this evidence shows.

5 Explain how ice cores are used to collect evidence of ancient levels of carbon dioxide.

6 The average pre-industrial level of CO_2 was 280 ppm. The current value is 400 ppm. Calculate the percentage **increase** in the concentration of CO_2

KEY INFORMATION

Remember you do not need to know the actual levels of carbon dioxide through the ages.

KEY CONCEPT

Greenhouse gases

Learning objectives:

- describe the greenhouse gases
- explain the greenhouse effect
- explain these processes as interaction of short and long wavelength radiation with matter.

KEY WORDS

wavelength
radiation
greenhouse
absorb

The average temperature of the Earth is 14 °C. This is because we have a blanket of gases as an atmosphere that protects us by keeping the temperatures relatively stable. The greenhouse effect is a natural phenomenon that is beneficial for us. If we are found to be altering it, then that is another story.

Why are greenhouse gases important?

Water vapour, carbon dioxide and methane are the greenhouse gases in our atmosphere. These gases allow **radiation** from the Sun to pass through and warm the Earth. We all feel this on a bright sunny day.

What is less obvious, but very important, is that the warm Earth radiates energy back out into the atmosphere and out into space, otherwise the Earth would get warmer and warmer.

The **greenhouse** gases 'trap' some of this radiation and help to keep us relatively warm.

So greenhouse gases in the atmosphere maintain the temperatures on Earth high enough to support life.

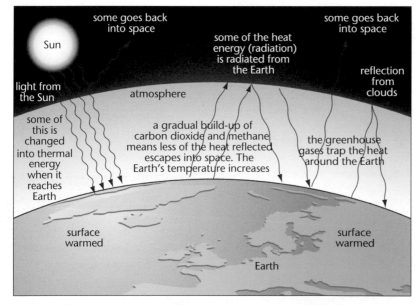

Figure 9.15 Solar radiation reaches Earth. Energy radiated back from the Earth is trapped by the greenhouse gases.

DID YOU KNOW?

The idea of a greenhouse effect was first put forward in 1824 by Joseph Fourier. It is estimated that the greenhouse effect makes the Earth 33 °C warmer than it would be otherwise. What is the temperature where you are today? Now reduce that by 33 °C ...

1 **Name two greenhouse gases.**

2 **Explain why greenhouse gases are important to life on Earth.**

How does the greenhouse effect work?

Radiation from the Sun is of short **wavelength**. It can pass through the atmosphere and not be **absorbed** by the atmospheric gases.

The gases allow *short wavelength* radiation to pass through the atmosphere to the Earth's surface. The gases then absorb the outgoing *long wavelength* radiation from the Earth, causing an increase in temperature. This temperature increase is beneficial to us as the temperature of the Earth is kept relatively stable.

Even though the temperature at the polar caps gets very low and the temperature at the equator can get very high, between these limits life can be sustained. Both our Moon and other planets experience temperatures way outside these limits. Life, as we know it, would be difficult to sustain elsewhere.

3 Describe the wavelength of incoming solar radiation compared with outgoing radiation from the Earth. Use a diagram to help.

4 Venus has an atmosphere of carbon dioxide. Suggest whether the average temperature will be higher or lower than expected when considering its distance from the Sun. Explain your reasoning.

> **DID YOU KNOW?**
>
> The surface temperature on the Moon varies between –173 °C and +127 °C. The range on Earth is –89 °C to +58 °C.

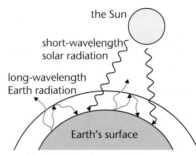

Figure 9.16 Short wavelength radiation from the Sun enters the atmosphere. Long wavelength radiation radiated back from the Earth is absorbed by the greenhouse gases.

What types of radiation are involved?

Radiation from the Sun is of short wavelength such as ultraviolet (UV), light and near infrared (near IR). It can pass through the atmosphere and not be absorbed by the gases. The radiation that comes back from the Earth is of a longer wavelength, which is far IR. This long wavelength IR radiation is absorbed by the molecules of water vapour, carbon dioxide and methane, which are the greenhouse gases.

The electromagnetic spectrum identifies radiation according to its wavelength and/or frequency.

This is an example of the spectrum to show the relative wavelengths of UV, visible light, near IR and far IR.

Figure 9.17 The relative wavelengths of UV, near IR and far IR, showing how incoming radiation is different from outgoing radiation

5 About 71 per cent of the total incoming solar energy is absorbed by the atmosphere or reaches Earth. Suggest what happens to rest of the solar energy.

6 Suggest what the Earth would be like if we did not have greenhouse gases.

> **REMEMBER!**
>
> You do not need to remember the names of the types of radiation.

Human activities

Learning objectives:

- describe two activities that increase the amounts of carbon dioxide and methane
- evaluate the quality of evidence in a report about global climate change
- recognise the importance of peer review of results and of communicating results to a wide range of audiences.

KEY WORDS

correlation
deforestation
peer review
speculation

The increasing level of carbon dioxide from the use of fossil fuels is now concerning world governments. This is particularly because the global population has risen so much in the last few decades, which means even more fuel is being used. However, other human activities are also producing higher levels of greenhouse gases. Do we know what their effect will be?

Human activities

We have seen that greenhouse gases are essential for maintaining stable temperatures on Earth so that life, as we know it, can be sustained. This is a natural phenomenon.

However, as far back as 1896 scientists were concerned that some human activities *increase* the amounts of greenhouse gases in the atmosphere and that we are altering the balance of the greenhouse effect. A Swedish chemist called Svante Arrhenius suggested that doubling the CO_2 level would cause the average temperature on Earth to increase by 5 °C.

The original concern was over the burning of coal, producing carbon dioxide. As more fossil fuels are burned, more carbon dioxide is produced. You can see this in the graph.

The increase in the percentage of carbon dioxide in the atmosphere over the last 100 years **correlates** with the increased use of fossil fuels.

There are now concerns over many other activities, that not only increase the amount of carbon dioxide but also increase the amount of methane. The human activities that affect these greenhouse gases include more:

- carbon dioxide because of
  1. combustion of fossil fuels
  2. **deforestation**
- methane because of
  1. more animal farming (through digestion and decomposition of waste)
  2. decomposition of rubbish in landfill sites

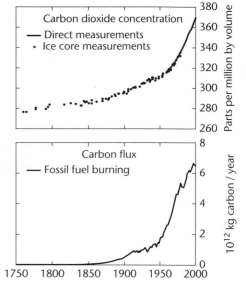

Figure 9.18 Graphs showing that fossil fuel use has risen and carbon dioxide levels have risen at the same time. They correlate.

1 Describe two reasons why carbon dioxide levels have increased in the last 200 years.

2 Explain why deforestation may lead to increased levels of carbon dioxide.

Evidence of human activity impact

We saw from ice core measurements of carbon dioxide that scientists believe that the amount of carbon dioxide in the atmosphere has increased significantly since the industrial revolution and the increased use of fossil fuels for transport, electricity production and machinery for manufacturing.

The evidence is **peer-reviewed** but questions asked include: 'Are these measurements accurate? Could the data be interpreted in a different way?' Most scientists agree that CO_2 levels have increased.

However, a few scientists would say that it is not humans that have caused this, but believe it is part of the Earth's longer, natural cycles.

It is a difficult debate, but based on peer-reviewed evidence, many scientists believe that human activities will cause the temperature of the Earth's atmosphere to increase at the surface and that this will result in global climate change.

3 Explain why it is important for scientific data and interpretation to be peer-reviewed.

4 Suggest what will happen in the next 50 years if the average temperature of the Earth goes up by 1.5 to 2 °C.

Modelling climate change

It is very difficult to model such complex systems as global climate change. This is why the Intergovernmental Panel on Climate Change (IPCC) was set up through the United Nations in 1988. This panel draws on evidence from hundreds of scientists who contribute to papers and peer-review each other's evidence. They try to model what might happen and give probabilities to future events. For example, they *predict* that oceans will warm and that it is *very likely* that Arctic ice cover will decrease.

Figure 9.19 Human activities that produce more carbon dioxide or methane

They produce Assessment Reports that can be read and evaluated. These reports are written by scientists not only for other scientists but also for policy makers. The scientists need to carefully explain their conclusions to people who may not know the technical language. In these reports they are writing for a different audience.

When the models are presented to the public they are simplified models. This means that **speculation** and opinions may be presented in the media that may be based on only parts of the evidence and part of the complex interpretations. At any stage of any explanation, there may also be a biased viewpoint.

5 There has been an increase in methane from agriculture over the last 50 years. Explain how you would search for evidence of its effect on the temperature of the Earth.

6 Explain why a simple model of predicted climate change seen on a 2-minute TV news item may not give quality information.

> **REMEMBER!**
>
> You do not need to know the models of climate change or the work of the IPCC but you do need to know how to evaluate the quality of evidence.

> **DID YOU KNOW?**
>
> Petrol, used in car engines, comes from a fossil fuel. It burns in oxygen to make carbon dioxide. Increased carbon dioxide levels have been linked with climate change.

Global climate change

Learning objectives:

- describe four potential effects of global climate change
- discuss the scale and risk of global climate change
- discuss the environmental implications of climate change.

KEY WORDS

average global
  temperature
distribution
food producing
  capacity
erosion

There are many different causes of climate change, most of which are natural. Examples are changes in radiation from the Sun, plate tectonic movements and volcanic eruptions. There are certain human activities that are thought to be causes of recent climate change. If humans cause a change, it is called an 'anthropogenic change'.

DID YOU KNOW?

The natural causes of increasing average global temperature are called 'forcings'.

The scale of global climate change

The major cause of global climate change is an increase in the **average global temperature**.

As we have seen before, the increase in the amounts of greenhouse gases is thought to cause an increase in average temperature.

Human activities producing greenhouse gases add to the natural causes of the rise in average temperature.

These activities are burning fossil fuels, deforestation, increased agriculture and rubbish decomposition.

The production of extra carbon dioxide and methane is thought to lead to increased global temperatures. In the media this is called 'global warming.'

The potential effects of global climate change by warming include:

- sea level rise, which may cause flooding and increased coastal **erosion**
- more frequent and severe storms
- changes in the amount, timing and distribution of rainfall.

1 Describe which human activities may add to 'global warming'.

2 Explain two things that might happen if global warming continues.

The risks of global climate change

Scientists are concerned that if the average global temperature rises by 2 °C there will be irreversible changes taking place caused by a change in climate. Climate change is not a short-term weather pattern change that follows cycles, but a long-term change in the whole weather system. A large number of scientists contribute evidence to the Intergovernmental Panel

Figure 9.20 Areas at sea level are more at risk of flooding as sea levels rise.

Figure 9.21 Storms become more severe and more frequent.

Figure 9.22 Distribution of rainfall may change.

on Climate Change (IPCC) so that the risks of climate change can be assessed. No government can do this on its own, as it is a global issue.

The scientists use a number of indicators to demonstrate how the climate is changing. The most sensitive indicator is the retreat of glaciers.

Glaciers are mountain reserves of fresh water that is frozen. The mass of water now held in glaciers continues to reduce.

3 **Explain the difference between a weather pattern and a climate change.**

4 **Suggest why it is important that glaciers remain high in mass.**

Figure 9.23 Glaciers are sources of frozen fresh water that are diminishing.

Environmental concerns of climate change

There will be big changes in the environment while global warming continues. The impact will affect different regions in different ways. In some regions the impacts are already changing lives, for example, for the people of the Arctic region where ice sheets are retreating.

The main effects are predicted to be:

- temperature stress for humans and wildlife (some areas will become too hot for people to make a living)
- water stress for humans and wildlife (fresh water supplies will reduce in some regions)
- changes in the **food producing capacity** of some regions (the production of wheat and maize has already been affected)
- changes to the **distribution** of wildlife species (migration patterns are changing).

Scientists are working on complex models to try to determine what may happen, as there are many interrelated factors to take into account.

5 **Explain how each of the potential social effects of climate change listed in this topic is related to the physical changes that are taking place.**

> **REMEMBER!**
> ....
> You are not expected to remember all the details that are involved in the complex issue of global climate change but you should aim to be able to discuss the possible causes and implications.

Carbon footprint and its reduction

Learning objectives:

- explain that the carbon footprint can be reduced by reducing emissions of carbon dioxide and methane
- describe how emissions of carbon dioxide can be reduced
- describe how emissions of methane can be reduced.

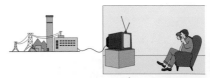

KEY WORDS

alternative energy
carbon capture
carbon off-setting
carbon neutrality

There are two responses to increasing average global temperature – reduce it or adapt to it. Adapting to it means coping with the social changes and ways of living. Reducing it means reducing greenhouse gases. The reduction of carbon dioxide and methane emissions is called a mitigation action so that we do not have to adapt.

Figure 9.24 We do not burn fuel when watching TV – or do we?

What is a carbon footprint?

The carbon footprint is the total amount of carbon dioxide and other greenhouse gases emitted over the full life cycle of a product, service or event.

We can often see what we are doing when burning a fuel that gives out CO_2. Examples are when we ride in a car or bus. At other times we may not notice. For instance, when we turn on the light or electric shower we do not always realise that the electricity we are using was probably generated by burning fossil fuels. We also do not take into account how much energy was used in making the car or the shower in the first place, and so how much carbon fuel was used to make it.

There is now an Intergovernmental Agreement, called the Kyoto Protocol, that, in essence, states that we should all reduce our carbon footprint.

Figure 9.25 Alternative energy sources: using solar energy

1 **Explain how playing a computer game has a carbon footprint.**

2 **Explain why governments agree that carbon footprints need reducing.**

Reducing the personal carbon footprint

Actions to reduce our personal carbon footprint include:

- increased use of **alternative energy** sources and hydrogen fuel cells
- energy conservation and energy efficiency in our homes
- energy efficiency by driving cars that use less fuel (i.e. have higher mpg figures).

3 **Explain how using solar energy reduces CO_2 emissions.**

4 **Explain whether using solar energy panels reduces the carbon footprint.**

Figure 9.26 Alternative energy sources: using nuclear energy

Figure 9.27 Alternative energy sources: using wind energy

5 Look at Figure 9.28. Explain why the methods with labels in blue are energy conservation methods and why the methods with labels in green are energy efficiency methods.

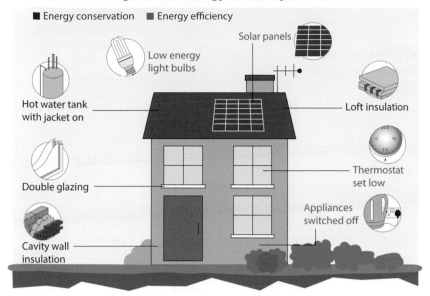

Figure 9.28 Using less energy by installing more insulation or using energy more efficiently

Governments reducing the carbon footprint

It is important that we reduce our own personal carbon footprint but there are some measures that governments or large companies or organisations will need to implement as we cannot do this on our own. These include:

- **Carbon capture** and storage – taking the emissions of CO_2 from a power station and depositing the CO_2 into an underground geological formation so that it will not enter the atmosphere. This is a new idea and so will need investment.
- Carbon taxes and licences – a system where a polluter pays a tax if they are emitting greenhouse gases. This is to give an incentive to the user to reduce CO_2 emissions. Another way is to give a licence or a permit to many users and put a cap on the total amount of CO_2 that can be emitted overall. The users then trade the permits amongst each other, so that some can emit more than others if they have bought their permits.
- **Carbon off-setting**, including increasing the carbon sink through tree planting and reforestation. Carbon off-setting can take place on a small local scale or a large scale such as the European Union Emission Trading scheme. This trading market grew out of the agreements reached through the Kyoto Protocol.
- **Carbon neutrality** – zero net release. This is a scheme where people and organisations aim to take out of the atmosphere as much CO_2 as they put into it, for example, by planting trees that use the equivalent CO_2 quantity as their emission from their electricity use. The idea is to produce a zero carbon footprint.

6 Explain two different ways by which forestation can help reduce the carbon footprint.

DID YOU KNOW?

This idea could have been tried by storing the CO_2 in the ocean, but this would have made the acidification of the oceans worse.

REMEMBER!

You do not need to know all the details of this topic but it is important to know where to find current information and be able to discuss it.

Limitations on carbon footprint reduction

KEY WORDS

economic
  considerations
international
  co-operation
population rise
scientific
  disagreement

Learning objectives:

- give reasons why actions to reduce levels of carbon dioxide and methane may be limited
- give reasons why methane is difficult to reduce.

We have seen that the reduction of carbon dioxide and methane emissions is called a mitigation action so that we do not have to adapt. We have also seen that although we can reduce our personal carbon footprint some actions need organisation by governments. Why does this not always work?

Difficulties in reducing methane levels

Methane is another greenhouse gas that has risen considerably in the last decades alongside the **rising population**. This is because as the population has risen:

- more grazing animals have been farmed to produce more meat, causing more methane waste. (This is an especially important factor where deforestation has also occurred to make way for grazing lands.)
- more wet cultivation fields for rice growing have been developed
- more rubbish landfill sites have opened.

These all cause methane levels to rise.

1. **Explain why the trend in population growth correlates with the trend in methane increase.**

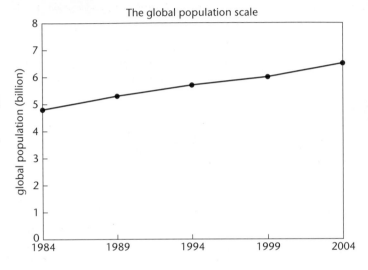

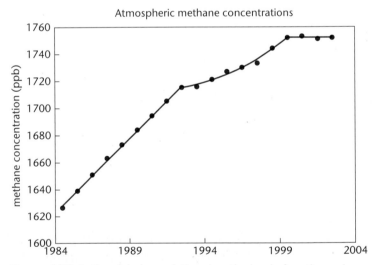

Figure 9.29 As the global population rises the level of methane rises.

Trying to reduce the carbon footprint

Although there is some agreement that the carbon footprint has to be reduced, it is often seen as someone else's problem. We do not always make a success of reduction because:

- Lifestyle changes are needed. Everybody needs to change their need for fuel in order to make a difference and it is not always easy or comfortable to do that. If everybody used a lot less fossil fuels or used different sources of energy, that

KEY SKILLS

When looking at two graph plots, look for pattern changes and correlations.

would make a huge difference, but it has to happen right around the world, particularly in the developed nations, especially those with big populations.

- There is a lack of public information and education. This happens because scientists do not always explain their evidence in a way that is easy to understand by people who do not have the technical language. Unfortunately, the scientific models are often complex and cannot be easily simplified. It also needs people to be more aware that they need to find out information and be pro-active in looking.

2 Suggest five steps people could take to change their lifestyle to reduce CO_2 emissions.

3 Suggest how public information could be improved so that evidence about global warming and the need for carbon footprint reduction could be debated more fully.

Other problems in reductions

There are also some major international problems with trying to reduce the carbon footprint which are not easily solved. These include:

- **Scientific disagreement** over the causes and consequences of global climate change. This disagreement is being challenged and reduced by the development of peer-reviewing of evidence by large numbers of scientists who are experts in a wide range of related fields, being drawn together by the IPCC. The bias of organisations who benefit from selling fossil fuels is also being scrutinised.
- **Economic considerations**. There is disagreement over a model that tries to predict the cost of trying to reduce the CO_2 concentration. The model is to reduce levels from 450 ppm, which governments have accepted as a target, to 350 ppm by changing the technology we use. This is because the lower figure is said to be the CO_2 level we should really be aiming for but it will cause instability. There is a wide range of opinions, including one that says there would be widespread unemployment and another that says in the long term there would be better employment prospects. Another says it will cost more whilst another says that emissions can be eliminated at no extra cost. This uncertainty does not help decision-making.
- Incomplete **international cooperation**. Not all countries agree with the vast majority who have signed agreements on carbon reduction. There are 192 signatories to the Kyoto Protocol, but Canada has withdrawn and the USA has not ratified its signature. 26 states have accepted the new Doha Amendment with targets for reduction up to 2020.

4 Suggest three reasons why information about the need to reduce the carbon footprint could be *biased*.

5 It is often difficult to decide the ownership of emissions contributing to a carbon footprint. For instance, a large sports stadium may claim that it has a very small carbon footprint when empty. Evaluate this claim.

DID YOU KNOW?

There is a target figure for the level of CO_2 that governments agree as a global target. It is measured in parts per million, ppm, not as a percentage.

Atmospheric pollutants from fuels

Learning objectives:

- describe how carbon monoxide, soot, sulfur dioxide and oxides of nitrogen are produced by burning fuels
- predict the products of combustion of a fuel knowing the composition of the fuel
- predict the products of combustion of a fuel knowing the conditions in which it is used.

KEY WORDS

....

sulfur dioxide
oxides of nitrogen
particulates
hydrocarbons

Burning a fuel produces energy, which is what we want. However, burning a fuel can also produce unwanted products. If a fuel contains sulfur, then acid rain can result. If a fuel is burned in a shortage of oxygen then soot (carbon particles) may be produced, which is dirty and bad for human health. Burning fuels is a major source of pollution in the atmosphere.

Figure 9.30 There are regulations in many urban areas on the types of fuels you can burn so that the air in cities is kept clean.

Common atmospheric pollutants

The combustion of fuels is a major source of atmospheric pollutants.

Most fuels, including coal, contain carbon and other elements such as hydrogen and may also contain some sulfur. The sulfur combines with oxygen to make **sulfur dioxide**.

When a fuel is burned the gases given off will include carbon dioxide and water vapour. However, they may also include unwanted gases such as carbon monoxide, sulfur dioxide and **oxides of nitrogen**.

Burning fuels may also release solid particles and unburned **hydrocarbons**. These form **particulates** in the atmosphere.

1 Write down three unwanted pollutants that are released when a fuel is burned.

2 Explain the term 'incomplete combustion'.

Gaseous pollutants

Carbon monoxide, sulfur dioxide and oxides of nitrogen are the gases that are pollutants.

DID YOU KNOW?

....

Carbon dioxide is not included in this group as, although it is a greenhouse gas and levels need to be reduced, it is not a harmful gas.

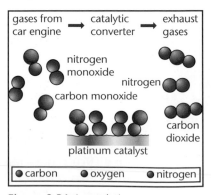

Figure 9.31 A catalytic converter

The environmental effects, how they are caused and how they can be removed or reduced are given in the table:

Gaseous pollutant	Carbon monoxide (CO)	Sulfur dioxide (SO₂)	Oxides of nitrogen (NOₓ)
how they are caused	incomplete combustion in a shortage of oxygen	burning coal or petrol containing sulfur in oxygen	from N₂ and O₂ at very high temperatures such as in an engine
how they can be reduced	converted to CO₂ by a catalytic converter	removed from power stations by capturing with limestone	converted to N₂ by a catalytic converter

$$S \quad + O_2 \quad \rightarrow \quad SO_2$$
$$C_7H_{16} + 8O_2 \quad \rightarrow \quad CO_2 + 6CO + 8H_2O \quad \text{Incomplete combustion}$$
$$C_7H_{16} + 11O_2 \quad \rightarrow \quad 7CO_2 + 8H_2O \qquad \text{Complete combustion}$$

Figure 9.32 Sulfur dioxide is removed and captured at source in this power station. Note that the clouds from the cooling towers are droplets of water not smoke.

REMEMBER!

You do not need to remember these equations but you do need to be able to predict the products of combustion.

3 Describe how some pollutants can be removed before they enter the atmosphere.

4 Use the equations above to explain why CO can be formed when fuels are burned.

5 The pollutant nitrogen monoxide is initially formed in engines. Give a balanced chemical equation for its formation.

Pollution by particulates

Solid particles and unburned hydrocarbons form particulate matter (PM) in the atmosphere, and are measured by their diameter. The size of the particles ranges in diameter and some are more penetrating in the lungs than others. Particles with a diameter of less than 10 micrometres (PM₁₀) can get past the filtering mechanism in the nose and go into the airways to the lungs. With diameters of less than 2.5 micrometres (PM₂.₅) they can penetrate the alveoli. These particles come from unburned fuel and cause health problems. They need to be reduced.

6 Look at Figure 9.33. Identify the three particulate contaminants with the smallest diameters.

7 Use Figure 9.33 to explain why pollution from vehicles is a problem.

8 Suggest how particulate emissions from cars might be reduced.

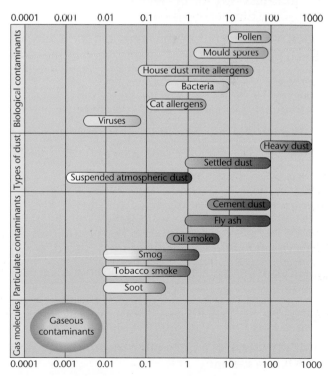

Figure 9.33 Diameters of airborne particles. The measurements are in micrometres (1000 micrometres = 1 mm).

Properties and effects of atmospheric pollutants

Learning objectives:

- describe and explain the problems caused by increased amounts of oxides of carbon, sulfur and nitrogen as pollutants in the air
- describe and explain the effects of acid rain
- evaluate the role of particulates in damaging human health.

KEY WORDS

acid rain
global dimming
particulates
toxicity

Gases from burning fossil fuels pollute the atmosphere, but that isn't the only problem. Solid particles can also be released, which are referred to as particulate matter (PM). However, is burning fossil fuels the only source of these PMs? Are there natural causes too?

Toxic gases

Carbon monoxide is a **toxic** gas made during the incomplete combustion of carbon. It is colourless and does not smell, and so is not easily detected. It is often called the silent killer as it makes people feel mildly unwell at first. Then they lose consciousness, without smelling any gas, and can die. It is particularly dangerous if you breathe it when you are asleep.

Carbon monoxide enters the lungs and combines with haemoglobin in the blood. It takes the place of oxygen in the red blood cells and so reduces the capacity of the blood to carry oxygen.

Figure 9.34 Regularly checking boiler emissions for carbon monoxide is important.

1. Suggest why carbon monoxide can cause heart problems.

2. Explain why it is so important to check appliances for carbon monoxide emissions.

Acid rain

Coal contains sulfur, so burning coal will make not only carbon dioxide but also sulfur dioxide. Petrol and diesel also contain sulfur but most is removed before it is sold.

Emissions from petrol or diesel burning in cars contain oxides of nitrogen.

Both sulfur dioxide and oxides of nitrogen dissolve in water causing **acid rain**. Acid rain damages plants and buildings.

Sulfur dioxide and oxides of nitrogen cause respiratory problems in humans.

This can also happen when oxides of nitrogen form photochemical smog, when sunlight interacts with the polluting gases.

DID YOU KNOW?

Nitrogen dioxide is NO_2 but there are other oxides of nitrogen in exhaust gases so they are grouped together as NO_x. The common name is then NOX gases.

Figure 9.35 Cars need to have an annual check for the levels of NO_x gases they emit.

3. What is the cause of acid rain?

4. Explain how acid rain causes damage.

Figure 9.36 Forests in the path of acid rain can be severely damaged if they are grown on certain types of soil.

Particulates

Particulates cause **global dimming**, reducing the amount of sunlight that reaches the Earth's surface. This can not only happen with burning fossil fuels but also through natural processes such as volcanic eruptions.

Figure 9.37 Dust and other particulates from natural causes or burning fossil fuels can cause global dimming.

KEY INFORMATION

Remember that rain is already slightly acidic as it dissolves CO_2 from the atmosphere. Sulfur dioxide and oxides of nitrogen make the rain *more* acidic, to levels where damage is done.

Particulates cause health problems for humans because of damage to the lungs. The size of the particles ranges in diameter and some are more penetrating in the lungs than others. Particles with diameters of less than 10 micrometres (PM_{10}) can go into the airways to the lungs. With diameters of less than 2.5 micrometres ($PM_{2.5}$) they can penetrate the alveoli. Some have different shapes and so they are just the right shape to penetrate and stay in lung tissue. Some are small enough to get into the blood stream. Particulates are deemed to be the most lethal pollution as they have the ability to go deep into lungs and enter the blood stream. They can lodge in the body causing heart attacks and DNA mutations. There is no safe level of particulate matter.

Figure 9.38 The damaging effects of acid rain

5 Use Figure 9.33 (topic 9.10) to estimate how many times larger setting dust is than soot from diesel cars.

6 Global dimming is caused by particulates in the Earth's atmosphere.

  a Suggest how global dimming may affect the Earth's climate and global warming.
  b Governments have put measures in place to reduce particulate emissions. Explain the impact this may have on the climate.

MATHS SKILLS

Use ratios, fractions and percentages

Learning objectives:

- use fractions and percentages to describe the composition of mixtures
- use ratios to determine the mass of products expected
- calculate percentage yields in chemical reactions.

To support life as we know it, there must be a proportion of oxygen in the atmosphere. The percentage of oxygen in the air was not always 20% and there was no oxygen until plants began to photosynthesise. How does a fifth of our atmosphere contain oxygen when animals are constantly using it? Why did the percentage composition of the atmosphere change?

The composition of air

Air is a mixture of gases. It contains nitrogen and oxygen and there are other gases present such as carbon dioxide, water vapour and noble gases like argon in small amounts.

If the other gases are taken out of the diagram it will show that the air is **approximately** $\frac{4}{5}$ nitrogen and $\frac{1}{5}$ oxygen. Approximately means 'almost'. If the circle is divided into 5 equal parts. 4 parts represent nitrogen and 1 part represents oxygen. The **fractions** are $\frac{4}{5}$ and $\frac{1}{5}$, respectively.

If the circle is now divided into 10 equal parts (so double the number of parts) then 8 represent nitrogen and 2 represent oxygen. The fractions are $\frac{8}{10}$ and $\frac{2}{10}$, respectively.

Now imagine that each of the ten **segments** was divided into 10. There would be 100 segments in the circle. 80 of these would represent nitrogen and 20 segments would represent oxygen. The fractions are $\frac{80}{100}$ and $\frac{20}{100}$, respectively.

Fractions representing $\frac{1}{100}$ are **percentages**. So $\frac{80}{100}$ is 80% and $\frac{20}{100}$ is 20%.

Now as fractions, percentages and a ratio the summary is:

$\frac{4}{5}$	$\frac{8}{10}$	$\frac{80}{100}$	80%	Ratio 4:1
$\frac{1}{5}$	$\frac{2}{10}$	$\frac{20}{100}$	20%	

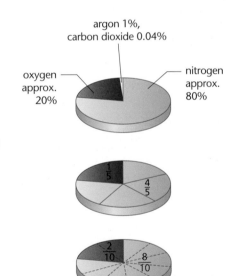

Figure 9.39 Pie charts showing the proportions of gases in the air

REMEMBER!

If you are drawing a pie chart, you draw the segments in order from largest to smallest clockwise.

1. The atmosphere of Mars is approximately $\frac{19}{20}$ CO_2, $\frac{1}{20}$ other gases. What is (a) the percentage of gases and (b) the ratio of gases?

2. If an ancient atmosphere had been 3:1 methane to ammonia, what would be the percentage of each gas?

Ratios of substances

In a chemical reaction an equation will tell us how many molecules will react in the process. Look at these two reactions:

$NaOH + HCl \rightarrow NaCl + H_2O$	$NaOH + H_2SO_4 \rightarrow Na_2SO_4 + H_2O$
	This is not balanced and mass is not conserved. It needs to be: $2NaOH + H_2SO_4 \rightarrow Na_2SO_4 + 2H_2O$
This is balanced and mass is conserved.	
Ratio is 1:1 [1 NaOH to 1 HCl]	Ratio is 2:1 [2 NaOH to 1 H_2SO_4]

Calculating the percentage yield

If there are two ways to make a substance it is really useful to know which method will produce the highest yield (amount of product) – but why use percentages and not fractions?

You may use one method where you need to filter a solution to get a solid product and lose a lot of it on the filter paper. You were expecting to make 12 g and only made 9 g. Your yield would be $\frac{9}{12}$ of what you were expecting. Another method may yield 20 g when you were expecting to achieve 24 g. Your yield would be $\frac{20}{24}$. How do you compare them easily? In this case, you can change the fraction to the same, lowest denominator – that is, to twelfths.

	Reaction 1 Method A	Reaction 1 Method B	Reaction 2
Yield	$\frac{9}{12}$	$\frac{20}{24}$	
Change to same lowest denominator	$\frac{9}{12}$	$\frac{10}{12}$	
You can see that method B produces more.	$\frac{9}{12}$	$\frac{10}{12}$	
What if it is not easy to change the fractions?	$\frac{9}{12}$	$\frac{10}{12}$	$\frac{5}{7}$
In order to compare the fractions, a common denominator is needed. In this case, the lowest common multiple is 84	$\frac{63}{84}$	$\frac{70}{84}$	$\frac{60}{84}$
Much easier to make all the fractions into percentages for comparison. Using a calculator, divide the numerator by the denominator and multiply the result by 100. Round your answer up or down to the nearest whole number for convenience in this case.	$\frac{9}{12} \times 100$ = 75%	$\frac{10}{12} \times 100$ = **83%**	$\frac{5}{7} \times 100$ = 71%

So if you want to make comparisons in yields you can use the formula:

$$\text{Percentage yield} = \frac{\text{Mass of product actually made} \times 100}{\text{Mass of product expected}}$$

3 What is the ratio of reactants in this reaction? $Mg + 2HCl \rightarrow MgCl_2 + H_2$

4 What is the ratio of reactants in this reaction? $3Fe + 2O_2 \rightarrow Fe_3O_4$

5 Determine the ratio of $H_2O:O_2$.

$4NH_3 + 5O_2 \rightarrow 4NO + 6H_2O$

6 N_2 reacts with O_2 in the ratio 1:2. A single product is formed. Give the balanced chemical equation for the reaction.

7 Calculate the percentage yield if you obtained 17.2 g and were expecting to get 26 g.

8 What mass did you make if your percentage yield was 80% and you expected 32 g?

9 0.58 g of product was made. 5.37 g of product was expected in theory. Work out the percentage yield.

10 27.0 g of product was produced. This was a yield of 31%. Calculate the mass of product expected.

DID YOU KNOW?

The percentage yield is a big consideration in making an industrial chemical.

Check your progress

You should be able to:

identify the gases of the atmosphere →	recall the proportions of gases →	explain how the balance of the gases is maintained
describe the ideas about the Earth's early atmosphere →	interpret evidence about the Earth's early atmosphere →	evaluate different theories about the Earth's early atmosphere
identify the processes allowing oxygen levels to increase →	explain the role of algae in the composition of the atmosphere →	recall the equation for photosynthesis
describe the main changes in the atmosphere over time →	describe some of the likely causes of these changes →	explain how ancient levels of carbon dioxide are measured
describe the greenhouse gases →	explain the greenhouse effect →	explain these processes as interaction of radiation with matter
evaluate the quality of evidence in a report about global climate change →	describe uncertainties in the evidence base →	recognise the importance of peer review of results and of communicating results to a wide range of audiences
discuss the scale of global climate change →	discuss the risk of climate change →	discuss the environmental implications of climate change
describe how emissions of carbon dioxide can be reduced →	describe how emissions of methane can be reduced →	give reasons why actions on reductions may be limited
describe some common unwanted products of combustion →	predict the products of combustion of a fuel knowing the composition of the fuel →	predict the products of combustion of a fuel knowing the conditions in which it is used

Worked example

The air contains a mixture of gases. One gas occupies 80% of the total.

1 **Identify the gas.**

    a oxygen   b carbon dioxide

    c (nitrogen)   d helium

> The answer nitrogen is correct.

2 **Explain which gases formed the early atmosphere and where they came from.**

They were carbon dioxide and nitrogen and they came from volcanoes.

> This answer is correct. The process is called degassing.

3 **Describe how in the early atmosphere:**
a levels of carbon dioxide were reduced
b levels of oxygen were increased.

(a) Plants used up the carbon dioxide.

(b) Plants gave out oxygen.

> The answers given are not incorrect but neither answer is detailed enough to pick up marks. Correct answers: (a) Plants used carbon dioxide for photosynthesis. Carbon dioxide was also used to make animal skeletons and shells, which formed sedimentary rocks. (b) Through photosynthesis plants produced oxygen. Over time levels increased.

4 **Explain how the levels of oxygen and carbon dioxide are maintained in the atmosphere today.**

Plants use CO_2 during photosynthesis and give out oxygen. Animals use oxygen for breathing in and breathe out CO_2. It's a cycle.

> The answer is correct for maintaining levels of oxygen. However, air is breathed in and out. Oxygen is used for respiration in cells, not for breathing.

5 **Write the symbol equation for photosynthesis.**

$CO_2 + H_2O \rightarrow C_6H_{12}O_6 + O_2$

> The student has recalled the reactants and products but has not understood how to balance the equation, which should be $6CO_2 + 6H_2O \rightarrow C_6H_{12}O_6 + 6O_2$

6 **Use the table and your knowledge to suggest what advice you would give to governments trying to reduce the total CO_2 emissions for their country.**

Fuel	Natural gas	Aviation fuel	Car petrol	Heating oil	Coal
CO_2 emitted in g for each unit of energy	50	66	68	69	91

They should reduce the amount of coal used first. Then they should tell people to use natural gas for heating instead of heating oil.

> The student has used data in the table to suggest two reductions. However, all the data shows a high level of CO_2 emission. The student should include suggestions such as the use of solar energy, lifestyle changes, reducing travel, reducing heating, increased insulation, etc.

End of chapter questions

Getting started

1 Identify the gas that causes acid rain.

    A sulfur dioxide   **B** hydrogen
    C carbon monoxide   **D** argon   `1 Mark`

2 Which of these gases is a greenhouse gas?

    A hydrogen   **B** oxygen   **C** argon   **D** methane   `1 Mark`

3 Suggest two factors that caused carbon dioxide levels to decrease in Earth's early atmosphere.   `2 Marks`

4 A species of bird has started to migrate to a different island further north than usual. What could this be due to?

    A increased oxygen levels   **B** global warming
    C acid rain   `1 Mark`

5 Work out the formula mass of sulfur dioxide SO_2. [A_r S is 32, A_r O is 16]   `1 Mark`

6 Peter is building his own new house. Suggest two things he could include in the building to reduce his carbon footprint.   `2 Marks`

7 Match the readings to the change.

ppm CO_2 27 31 35 37		increase in global warming
ppm NO_x 3.6 4.4 5.2 5.6		increase in greenhouse gas
°C 12 12.5 12.9 13.1		increase in acid rain

`2 Marks`

Going further

8 Identify the atmospheric pollutants given off from the incomplete combustion of fuels.   `1 Mark`

9 Compare the wavelength of radiation entering the Earth's atmosphere to that being given out at the Earth's surface.   `1 Mark`

10 Describe how respiration increases the level of carbon dioxide.   `2 Marks`

11 Emily

    a has a large house
    b has a large garden with grass
    c has a car with a large engine
    d flies on holiday five times a year.

    Describe one way that Emily could reduce her carbon footprint in each case.   `4 Marks`

12 These average winter temperatures at a glacier have changed each year.

    Data table for Years 1–7:   `2 Marks`

−2.0°C	−1.8°C	−2.5°C	−1.5°C	−1.5°C	−1.1°C	−0.8°C

    Explain what may happen in 10 years' time.

More challenging

13 Write the full balanced equation for respiration.

`1 Mark`

14 Describe what would happen to the mean global temperature if there were no greenhouse gases.

`1 Mark`

15 Give one reason why not all the radiation from the Sun reaches the Earth's surface.

`2 Marks`

16 Explain why the reduction of cattle grazing and rice fields may limit global warming.

`2 Marks`

17 The amounts, in millions of tonnes, of NO_2 and SO_2 released in the UK every five years from 1970 to 2010, are shown in the table.

Year	1970	1975	1980	1985	1990	1995	2000	2005	2010
NO_2	2.67	2.57	2.67	2.60	2.88	2.31	1.79	1.58	1.12
SO_2	6.32	5.17	4.76	3.68	3.68	2.36	1.22	0.71	0.43

a Describe and explain the trend for NO_2 and SO_2.
b Suggest the impact on the environment.

`4 Marks`

Most demanding

18 The lifetime of methane in the atmosphere is much shorter than carbon dioxide although it has a greater global warming potential. Explain how these factors might affect global warming.

`2 Marks`

19 When burned in oxygen, pentane, C_5H_{12}, produces CO_2 and water.

a Write a balanced symbol equation of the reaction.
b Calculate the formula masses of pentane and carbon dioxide.
c Calculate the mass of CO_2 that 144 g of pentane produces.

`4 Marks`

20 Scientists believe that the Earth's early atmosphere was probably similar to Venus's current atmosphere.

	Earth's atmosphere / %	Venus's atmosphere / %
Nitrogen	78.1	3.5
Oxygen	21.0	trace
Carbon dioxide	0.039	96.5

a Explain why planets such as Venus are used as a model for the early atmosphere on Earth.
b Work out how many times more carbon dioxide there is on Venus than on Earth.
c Explain the difference in oxygen concentration.

`4 Marks`

`Total: 40 Marks`

SUSTAINABLE DEVELOPMENT

USING NATURAL RESOURCES

- Wood, clay, limestone and metal are used to make shelters.
- Wood, coal, gas and oil are used for warmth and transport.
- Wood can be grown again but coal and crude oil are finite.

THE WATER CYCLE

- Water evaporates from oceans and seas to form clouds.
- Clouds move the condensed water inland where it precipitates.
- Rain fills the rivers, lakes and reservoirs to provide fresh water.

REUSING AND RECYCLING

- More household waste is recycled to reduce the landfill sites.
- Bottles and metal cans are collected separately for recycling.
- Houses are better insulated to cut down energy needs.

NATURAL AND SYNTHETIC MATERIALS

- Natural and synthetic materials are used to make products.
- Natural materials include cotton, wood, wool and sand.
- Synthetic materials include glass and plastic.

METALS AS RESOURCES

- Copper and iron are useful metals needed on a large scale.
- Copper is used for electrical wiring and saucepans.
- Iron is used to make steel for bridges and buildings.

IN THIS CHAPTER YOU WILL FIND OUT ABOUT:

HOW CAN WE SUSTAIN RESOURCES FOR FUTURE GENERATIONS?

- Natural resources and agriculture provide food and fuel.
- Some resources can be sustainably developed for the future.
- Resources such as fossil fuels are finite and need managing.

HOW CAN WE ENSURE ACCESS TO DRINKING WATER AND TREATMENTS?

- Water needs to be free from microbes and solids.
- There is limited fresh water so seawater needs desalination.
- Sewage and waste water need treatment before recycling.

REDUCING RESOURCE WASTE

- Recycling reduces use of finite resources, saving for the future.
- Environmental impact is lessened by reducing energy consumption.
- Reusing materials saves valuable resources.

WHAT ARE LIFE CYCLE ASSESSMENTS?

- Life cycle assessments estimate environmental impact.
- Each stage of a product's life is assessed.
- Processing the materials to final disposal is assessed.

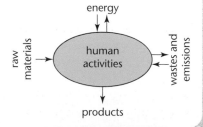

WHAT OTHER WAYS CAN WE EXTRACT METALS?

- Traditional mining needs large scale disposal of rock.
- Phytomining uses plants to extract metals.
- Bioleaching uses bacteria to leach metals.

KEY CONCEPT

Using the Earth's resources and sustainable development

Learning objectives:

- give examples of natural products replaced by synthetics
- give examples of products replaced by agricultural products
- distinguish between finite and renewable resources.

KEY WORDS

resources
sustainable
future generations
finite

Many of the things that we want or take for granted now are developed through advancing technology, using materials from finite resources. Can we keep developing this way and still leave enough for future generations? Are we damaging our environment so much that it will not sustain us in the future?

Providing for our needs

Humans have always used the Earth's **resources** to provide warmth, shelter, food and transport. Wood, clay and stone were the first materials to be used to make permanent structures. These materials were followed by copper, bronze and iron. At that time there were enough natural resources to satisfy the needs of the small population and leave enough for **future generations**.

To make our ways of life **sustainable** we need to be able to distinguish between natural resources, sustainable resources and **finite** resources.

Natural resources such as wood, stone and clay, supplemented by agriculture for cotton, wool and rice, provide food, timber, clothing and fuels. The wood needs to be managed though. Forests need to be replanted if logging removes them. A forest is a natural resource, a managed forest is a sustainable resource.

If we can use mostly sustainable resources there will be enough for the future. If we need to use finite resources then we need to use them sparingly (reduce the need), reuse them and recycle them, so that they can be used by future generations too.

Figure 10.1 New materials are needed for buildings, transport and warmth.

DID YOU KNOW?

Since the Industrial revolution we have changed the way we use resources for our necessities.

With an ever-growing population some of these resources are running out and there will be none for future generations.

REMEMBER!

Plastics are made from crude oil, which is a finite resource.

1 **Describe one object that used to be made of wood that is now made of plastic.**

2 **Describe one object that used to be made of cotton that is now made of plastic.**

Finite resources

A *finite resource* is one that cannot be made again, such as crude oil, coal and iron ore.

Finite resources from the ground, oceans and atmosphere are processed to provide energy and materials. Some of the *reserves* of these finite materials are running out, such as crude oil; some are *abundant*, such as sand.

Chemistry plays an important role in improving the industrial and the agricultural processes to provide these new products. Chemistry also plays an important role in ensuring *sustainable development*, which is 'development that meets the needs of current generations without compromising the ability of future generations to meet their own needs.'

Figure 10.2 Chemistry develops the materials needed for new products.

3 Explain the idea of a finite resource.

4 Suggest whether (a) sand (b) water (c) coal and (d) wood are renewable or finite resources. Explain your reasons.

Using data

When we are looking at how much natural or finite resources we are using it is important to monitor how much we are taking and using compared with other countries and how much we all reuse and recycle. We can monitor this by country or by person (per capita). For example, these are figures monitoring the UK use of building materials.

Resource	Cement materials	Brick fireclay	Gypsum	Slate	Building stone	Aggregates
thousand tonnes	17 390	7480	1700	750	690	238 700

Another time we can use data is when we are choosing which pathway to use when we are making chemicals for making new products. We need to use data such as atom economy, yield, rate, equilibrium position and usefulness of by-products.

5 Explain why it is important that the value for natural slate in the table above is kept lower than the value for aggregates.

6 One of the steps in making cement is to form calcium oxide by heating calcium carbonate to high temperatures for long periods.

$$CaCO_3 \rightarrow CaO + CO_2$$

Give one reason why this may not be sustainable and one reason it has an environmental impact.

Figure 10.3 Recycling is a key factor in sustainable development.

Potable water

Learning objectives:

- distinguish between potable water and pure water
- describe the differences in treatment of groundwater and salty water
- give reasons for the steps used to produce potable water.

Safe drinking water is essential for life. Where does it come from and how is it made safe? For humans, the appropriate quality drinking water should have sufficiently low levels of dissolved salts and microbes. How do you get safe water if it does not rain?

Producing potable water

Water that is safe to drink is called **potable** water. In a chemical sense, potable water is *not pure* water because it contains *dissolved substances*, which are needed by the body.

There are several methods used to produce potable water. The method chosen depends on available supplies of water and local conditions.

Traditionally, settlements grew up on the sides of river banks or lakes as the source of water was *abundant*. Other settlements relied on water from *groundwater* that collected in *aquifers*. They extracted it from *wells*.

In the United Kingdom (UK), as the population grew in cities, *reservoirs* were built to provide a constant supply of fresh drinking water for everyone.

The water is provided by *rain*. This water, with low levels of *dissolved substances (fresh water)* collects in the ground and in lakes and rivers and is then treated to make it potable.

There are three main stages in water treatment:

- **sedimentation** of particles so that solids drop to the bottom
- *filtration* of very fine particles using sand
- *sterilising* to kill microbes. Sterilising agents include chlorine, ozone or ultraviolet light.

1 **Explain the stages of how rain becomes drinking water.**

2 **Suggest why the sterilisation stage of water purification is carried out last.**

3 **Some data for a water sample is shown below. Explain whether the water is potable.**

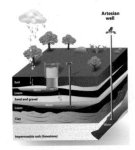

Figure 10.4 Groundwater is held in aquifers and is reached from the surface through artesian wells.

Figure 10.5 Huge reserves of water are held to provide drinking water.

Figure 10.6 A sedimentation tank at a water-treatment works.

	Bacteria per 100 ml	Lead / micrograms µ per ml	Nitrate / mg per l
Sample of water	2	2	15
Maximum allowed	0	10	50

Potable water from seawater

Seawater has so many substances dissolved in it that it is undrinkable. However, if supplies of fresh water are limited, **desalination** of salty water or seawater is needed to remove the dissolved substances.

Desalination can be done by *distillation* or by processes that use *membranes* such as **reverse osmosis**. These processes require large amounts of energy, so desalination is very expensive. It is only used when there is not enough fresh water.

Many countries with little fresh water rely on desalination processes using distillation or reverse osmosis. Spain is the largest operator of desalination plants in Europe.

4 Suggest why Spain operates desalination plants whereas the UK does not.

5 Explain which stage in distillation makes the process so costly.

Potable water for all

Water is a *renewable* resource. However, that does not mean that the supply is endless. If there is not enough rain in the winter, reservoirs do not fill up properly for the rest of the year. In the UK today, more and more homes are being built, which increases the demand for water. Industry and agriculture also use huge amounts. Producing potable water does have costs. It takes energy to pump and to purify it. If this energy is sourced from fossil fuels this adds to the increases in greenhouse gases and in turn to increases in global warming.

This then causes stress to *glaciers* that are a source of river production that provides us with fresh water. To minimise that stress we need to consider water *conservation*.

6 Suggest three ways in which domestic water can be conserved.

7 Suggest three ways in which industrial or agricultural water can be conserved.

8 The bar chart shows water consumption per capita for various regions per day.

a Estimate the ratio of water consumption between North America and sub-Saharan Africa.

b Suggest why sub-Saharan Africa has such a low consumption of water per capita.

Figure 10.7 A Spanish desalination plant producing potable water using reverse osmosis.

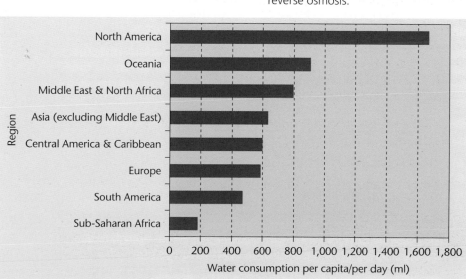

Bar chart: Water consumption per capita/per day (ml). Regions (y-axis) versus water consumption (x-axis, 0 to 1,800):
- North America: ~1,700
- Oceania: ~900
- Middle East & North Africa: ~800
- Asia (excluding Middle East): ~650
- Central America & Caribbean: ~600
- Europe: ~600
- South America: ~450
- Sub-Saharan Africa: ~200

REQUIRED PRACTICAL

Analysis and purification of water samples from different sources, including pH, dissolved solids and distillation

KEY WORDS

distillation
purity
boiling point

Learning objectives:

- describe how safety is managed, apparatus is used and accurate measurements are made
- recognise when sampling techniques need to be used and made representative
- evaluate methods and suggest possible improvements and further investigations.

Fresh drinking water is essential for everyone. Water needs to be made safe, which means testing and treatment. River water can be filtered and bacteria removed or seawater can be distilled.

❶ These pages are designed to help you think about aspects of the investigation rather than to guide you through it step by step.

Analysing water samples

Different skills are needed to analyse water and carry out a **distillation**, including applying sampling techniques, manual dexterity and the safe use of heating devices.

Planning your sampling

One way of selecting samples is by layered or section sampling. If 30 samples of river water are to be tested, they must be taken equally in each section over the whole stretch.

Imagine a new town was being planned which needs potable water. A river runs nearby. Your job is to test the water to find the best place to site a water treatment plant. The pH needs to be 6.5–9.5

The river section is 1 km long. It can be divided into three sections:

A. Open countryside over chalk   (0.5 km)
B. Past a large dairy farm   (0.2 km)
C. Past an old lead mine   (0.3 km)

Think about these questions:

❶ 30 samples of river water are tested altogether. Calculate the number of samples needed from each river section.

❷ Explain how you will test the pH of the samples.

DID YOU KNOW?

There are two steps to getting the most accurate results of chemical analysis:
- collecting samples and
- analysing samples.
Errors need to be reduced in both steps.

3 Other than 'What is the pH?', which other questions will you ask about the sites that the samples are taken from?

Some countries do not have access to fresh water, only seawater. This has dissolved salts in it that need to be removed. This is done by distillation. Figure 10.8 shows the apparatus for the standard technique.

Think about these questions:

4 Explain why the water in the flask is heated.

5 Why is a condenser attached to the arm of the heating flask?

6 Describe how distillation works.

7 Explain the purpose of the thermometer.

8 Why does the salt not evaporate?

9 Explain why the cold water flows into the *bottom* of the condenser and out into the sink from the *top* of the condenser.

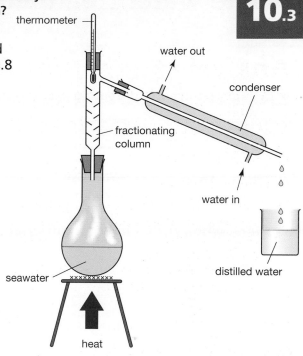

Figure 10.8

Analysing the results

When three samples are gathered, they are tested for **purity** by finding the **boiling point**. Jo and Akira tested their three samples:

	1st sample	2nd sample	3rd sample
Boiling point before distillation in °C	102.1	102.0	101.9
Boiling point after distillation in °C	101.4	101.6	100.8

10 Explain the results.

The way the equipment was used, and how techniques can be improved, should be considered.

11 Suggest ways to improve the sampling of river water above. For example, after answering question 3, would all the sites need to be included in the sampling?

12 Suggest ways to improve the measurement of pH or the distillation technique shown.

13 Suggest what other investigations could be done to analyse drinking water samples.

14 Taking averages is common in science.

a Calculate an average of the three sample boiling points both before and after the distillation.

b Explain whether separate or average boiling points are more useful.

c The pH values of 50 samples of river water at different locations were measured. Suggest whether an average or separate values would be more useful.

> **REMEMBER!**
>
> The water in the cooler jacket of the condenser does not mix with the water going through the centre of the condenser.

> **KEY INFORMATION**
>
> A simplified equation for section sampling could be: Sample size for each section = size of whole sample × size of section

Waste water treatment

Learning objectives:

- explain how waste water is treated
- describe how sewage is treated
- compare the ease of treating waste, ground and salt water.

Water is a precious commodity that we tend to take for granted in the UK as we have so much rain. The natural water cycle works well for us most of the time, but what happens when urban living, industrial development, intensive agriculture or dry weather put pressure on this natural resource?

Figure 10.9 Sewage pouring into a river. In the UK, sewage is treated to make sure it doesn't contaminate the waterways.

Water cycles

There is no new water on the planet; it is ancient water that is continually recycled.

The water cycle is a natural cycle, in which water from the oceans evaporates, condenses to clouds, moves over high ground and precipitates water back to the ground to rivers, lakes and aquifers.

For thousands of years we have lived with this natural cycle, drinking water from wells, rivers and lakes and excreting waste into the ground where the solids are broken down by bacteria and the water finds its way back to groundwater or rivers or evaporates to form clouds.

With growing populations, ways of treating **sewage** are needed as well as ways of providing clean fresh water to everyone.

Systems for dealing with sewage depend on the density of the population. In rural areas, each household has their own individual sewage system or on-site sewage facility. This is called a *septic tank*.

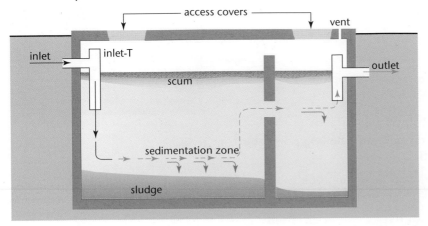

Figure 10.10 How a septic tank works

DID YOU KNOW?

The Great Stink of 1858 galvanised the Members of Parliament into action to pass laws allowing new sewage systems to be developed that did not put untreated sewage straight into the River Thames, literally straight under their noses: Go to www.skepticalscience.com and search for 'changing minds the great stink'.

Septic tanks allow *anaerobic* bacteria to develop that treat the sewage by decomposing it. The treated water leaks out through finger drains on to the land and the remaining sludge collects in the tank. Periodically the tank needs to be emptied.

1 **Describe how water from the sea ends up in aquifers.**

2 **Explain why some rural communities need septic tanks.**

The urban water cycle

Urban lifestyles and industrial processes produce large amounts of waste water that require treatment before being released into the environment. This can be from domestic washing machines, dishwashers and showers. It can be from industrial processes such as cleaning cycles and solvent usage in manufacturing processes and from cooling systems.

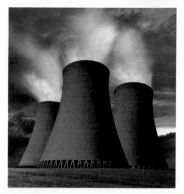

Figure 10.11 Industrial use of water is increasing.

The *industrial waste water* may contain pollutants such as organic matter and harmful chemicals that require extra treatment before it can go back into the fresh water cycle.

Agriculture is by far the biggest user of water. *Agricultural waste water* can cause problems to ecosystems by having too much nutrient in it. This can lead to *eutrophication*.

This is why sewage and agricultural waste water require treatment to ensure the removal of organic matter and harmful microbes.

Sewage treatment includes:

- *screening* and grit removal
- **sedimentation** to produce sewage sludge and *effluent*
- *anaerobic digestion* of sewage sludge
- **aerobic** *biological* treatment of effluent.

DID YOU KNOW?

Clean water saves more lives than medicines do. That is why, after disasters and issues in developing countries, relief organisations concentrate on providing clean water supplies.

3 **Suggest how sedimentation works and predict one limitation.**

4 **Explain why urban lifestyles need organised waste water management.**

The water footprint

Just as there is the monitoring of the carbon footprint, there is an organisation encouraging the monitoring of the water footprint.

International bodies describe water in categories. Green water is water from precipitation, blue water is in rivers, lakes and aquifers and grey water is waste water from agricultural, industrial and domestic use. It is grey water that needs treatment to go back into the water cycle.

5 **Suggest why Australia may be most affected by changes in water footprint.**

6 **Compare the relative ease of obtaining potable water from groundwater, saltwater and waste water.**

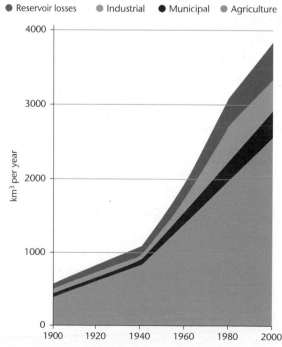

● Reservoir losses ● Industrial ● Municipal ● Agriculture

Figure 10.12 Graph of water usage between 1900 and 2000

Alternative methods of metal extraction

Learning objectives:

- describe the process of phytomining
- describe the process of bioleaching
- evaluate alternative biological methods of metal extraction.

KEY WORDS

phytomining
bioleaching
hyperaccumulators
toxic metals

Traditional mining means digging out huge chunks of rock and transporting it around. A new way of extracting some metals from these ores is to use bacteria. Is this feasible? What if we could clean our discarded sites of the toxic metals left behind? What if we could use plants to do this as well?

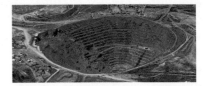

Figure 10.13 Traditional forms of mining need huge movements of rock.

HIGHER TIER ONLY

Phytomining

The Earth's resources of metal ores are limited. Copper ores are becoming scarce and the price of this valuable metal is rising sharply. Copper's electrical conductivity is high, so we use a lot of it for wiring and also in alloys. We do recycle it but we can also extract smaller amounts using other methods.

The new ways of *extracting copper* from *low-grade* ores include **phytomining** and **bioleaching**. These methods avoid the traditional mining methods of digging, moving and disposing of large amounts of rock.

Phytomining uses *plants* to absorb metal compounds. The plants are harvested and then burned to produce ash that contains the metal compounds. Plants will naturally absorb compounds from the soil through their root system. We need some of these chemicals for our good health. There are certain metals that plants will absorb that are poisonous to them and they will die. But there are other plants that will collect these metals in their leaves.

These plants are called **hyperaccumulators**. They concentrate the **toxic metals** in their tissue.

Nickel is very difficult to extract. We use it a lot, in stainless steel, alloys for coins and in batteries. However, it can be absorbed and stored by a particular *hyperaccumulator*.

If a hyperaccumulator could be planted in soil that had nickel in it and grown so that the metal was absorbed, then it could be harvested and the metal extracted from its leaves.

Figure 10.14 Some special plants can be used in phytomining.

Figure 10.15 Copper compounds extracted by bioleaching can be purified by electrolysis.

1. Explain three reasons why it is important that new ways of extracting copper ore are developed.

2. Explain why plants are important in the technique of phytomining.

Bioleaching

Bioleaching uses *bacteria* to produce *leachate* solutions that contain metal compounds. These can then be processed to obtain the metal. For example, copper compounds in leachate solution can be used to obtain pure copper by two methods:

- displacement using scrap iron
- electrolysis.

Bioleaching is a cleaner process than the traditional leaching that uses cyanide.

3. Explain how scrap iron and bacteria can be used to extract copper from waste copper ore.

4. Bioleaching is a slow process and a by-product is sulfuric acid. Explain how this might influence whether or not bioleaching is used.

Evaluating production

Some bacteria can convert metal sulfides to metal sulfates, which is a technique used for recovering mainly copper. However, the costs are still quite high and do not compare with smelting processes yet. So, although there may be some environmental advantages, the process will also need to have economic advantages too before it is used on a large scale in remote areas.

However, bioleaching could be used for cleaning up toxic metals from old industrial sites that have been contaminated so that new building sites can be claimed back from 'brown sites', instead of using green fields. This would help the sustainability cycle.

5. Suggest why mining and smelting may still be a more cost-effective way of producing copper.

6. The annual world production of copper is 1.9×10^7 tonnes. About 20% of this comes from bioleaching. Bioleaching is up to 90% efficient. Calculate the mass of copper produced annually by bioleaching.

DID YOU KNOW?

Metals such as iron and copper that have been mined since ancient times leached into the river in Spain that became known as Rio Tinto.

Figure 10.16 The Rio Tinto river

KEY SKILLS

You will need to know how to read data tables that will not always give a clear indication of the best method of metal extraction to use.

Life Cycle Assessment and recycling

Learning objectives:

- describe the components of a Life Cycle Assessment (LCA)
- interpret LCAs of materials or products from information
- carry out a simple comparative LCA for shopping bags.

KEY WORDS

Life Cycle
  Assessment
extracting
manufacture
disposal

We all want a clean environment but we also want the latest technology and transport. Each time a new product is advertised we look at the cost of it, but do we look at the cost of making it and the effects of disposing of it when we are done with it?

Life cycle assessments

A **Life Cycle Assessment (LCA)** is carried out to assess the environmental impact of a product in each of these stages of its life. It is sometimes called a 'cradle to grave' analysis. It not only considers the useful life of a product but also the materials and energy it takes to make the product and what happens to it on its **disposal**.

The stages in a simplified LCA are to consider:

- **extracting** and *processing* the *raw materials*
- **manufacturing** the product and *packaging*
- the *use* and *operation* of the product during its lifetime
- *disposal* at the end of its useful life.

Transport and distribution and *waste* are included at each stage. Let's look at shopping bags made of plastic and paper.

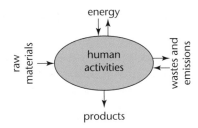

Figure 10.17 A Life Cycle Assessment diagram for a product

Stage	Plastic	Paper
Raw materials	Processing crude oil	Made from wood
Manufacturing	Do both use the same processes?	
Use and operation	Which lasts longer?	Which is stronger?
Disposal	How do you recycle, reuse or dispose of plastic or paper?	
Transport	Is this the same for both?	
Emissions and waste	Do the manufacturing processes give off emissions? What happens to waste plastic?	Do the manufacturing processes give off emissions? What happens to waste paper?

Figure 10.18 Comparing the LCAs of a plastic bag and a paper bag

There are a number of processes in this comparison that are the same. You need to identify the processes and outputs that are different and compare these. There is quite a lot of difference in the raw materials used and the disposal in this example.

1. Suggest which raw material is the most sustainable in this example.

2. Explain what happens at the end of the useful life of each of these bags. Suggest which one is easier to dispose of.

3. Suggest what the term 'cradle-to-cradle Life Cycle Assessment' means.

Objectivity and subjectivity

This exercise is done partly as a *description* and partly with numerical data. A description is called *qualitative* information.

Use of water, resources, energy sources used (*consumption*) and *production* of some wastes can be fairly easily quantified (put into *numerical* terms). This is called *quantitative* data.

It is much less easy to allocate numerical values to pollutant effects and requires *value judgements*. These are *subjective* (swayed by the person's viewpoint). So making LCAs is not a purely *objective* process (one where the data can only be interpreted in one way).

4. Identify whether the following statements are qualitative, quantitative, subjective or objective.

   a Plastic bags are more environmentally friendly than paper bags.

   b It takes 0.48 MJ of energy to make a plastic bag and 1.6 MJ to make a paper bag.

5. Two students have different viewpoints on how much CO_2 goes into the air from the making of paper or plastic. They can find no data on it. Explain why their LCAs may be different.

Advertising

It is important to identify the source of the LCA. It could be unbiased from an impartial agency, a reputable company or a protesting group that is fair. It *may* have some bias as a company may want to sell a lot of their product or because a group does not want the product to go ahead. LCAs can be devised that are *selective* or *abbreviated* when evaluating a product and be misused to reach pre-determined conclusions, for example in support of claims for advertising purposes. It is important to be critically aware of bias or non-bias.

6. Draw up a table for the LCA for bag use as on the previous page. Answer the questions in the table.

7. It takes an average of 119 000 MJ of energy to make a typical car.

   a Suggest how this sort of data can affect the accuracy of an LCA.

   b State one factor that is not taken into account in an LCA and how it might affect the manufacturer and consumer.

> **REMEMBER!**
>
> You only need to be able to compare the impact on the environment of the stages in the life of a product by description.

Figure 10.19 Considering objective data

> **DID YOU KNOW?**
>
> You only need to be able to use quantified data for energy, water, resources and wastes when it is readily available.

'Why wash china plates when you could have a clean plastic plate every day?'

Figure 10.20 Have they done an LCA? Are we persuaded?

Ways of reducing the use of resources

Learning objectives:

- describe ways of recycling and reusing materials
- explain why recycling, reusing and reducing are needed
- evaluate ways of reducing the use of limited resources.

KEY WORDS

recycling
limited resource
reduction of use
reuse of
  resources

Reduce, reuse, recycle. Is that what we should be doing? The population is growing and we have limited resources on the planet for everyone, both now and for future generations. What will it be like for our grandchildren if we take all the finite resources? What can we do about it?

Figure 10.21 Building materials are made from resources that often have to be mined or quarried.

Recycling and reusing

Metals, glass, building materials and clay ceramics are produced from limited raw materials that mostly have to be dug up from the ground.

Obtaining raw materials from the Earth by *quarrying* and *mining* causes *environmental impact*.

Some products, such as glass bottles, can be **reused**.

Figure 10.22 Reusing or recycling materials can cut down the need for quarrying or mining which have large environmental impacts.

Other products cannot be *reused* and so are *recycled* for a different use.

Glass bottles can also be recycled. They can be crushed and melted to make different glass products.

The amount of separation required for **recycling** depends on the material and the properties required of the final product.

Metals can be recycled by melting and re-casting or re-forming into different products.

For example, some scrap steel can be added to iron from a blast furnace to reduce the amount of iron that needs to be extracted from iron ore.

1. Describe two environmental impacts that a quarry has on local residents.

2. Explain why it is important to recycle metals and glass.

3. Some plastic bags are strong and can be repeatedly reused. Suggest why it is better to reuse than recycle.

Reducing the use of finite resources

Most plastics are made from crude oil, which is a **limited**, finite **resource**. It is not as easy to recycle as it is to recycle metal or glass.

Plastics do not degrade easily and take up valuable land when dumped in landfill sites that need to last for decades.

Some plastics are used in small quantities and greatly enhance our lives, but do we really need to use plastic bags?

The same limited resource – crude oil – is needed to produce the *energy* used in the *extraction of resources*, such as metal ores, and in the *manufacturing processes* of materials such as metal and glass, and objects such as cars and plastic furniture.

4. Explain why it is important to use crude oil wisely.

5. Suggest objects made from plastic that could be made from materials that will degrade.

6. It is common to see the slogan 'Reduce, reuse, recycle'. Discuss how this slogan can be applied to a typical car.

End use reduction

We have seen that the reuse and recycling of materials *reduces*:

- the use of limited resources
- use of energy sources
- waste
- the environmental impact.

We could also do one more thing. That is not to use so much in the first place.

The **reduction of use**, by *end users*, would have a massive impact on the amount of resources that we need to take out of the Earth at the start of the process of manufacturing.

7. It is possible to assemble cars a long distance from or near to the point where they will be sold (the market). Explain which way may save more limited resources.

8. An industry produces metal parts for precision hospital equipment. They are discussing how to set up an assembly line so that waste is reduced. Explain two things they could do.

9. The annual use of plastic carrier bags in the UK was 7.6×10^9 (7.6 billion). On the introduction of a charge, use fell by 80%.

   a Describe the environmental impact of the reduction.

   b Calculate the number of plastic bags used after the reduction.

DID YOU KNOW?

Milk used to be distributed in glass bottles that could be reused then recycled.

Figure 10.23 Metals can be re-melted with metal from new ore to reduce the amount of new ore needed.

Figure 10.24 Discarded plastic takes up valuable land and does not degrade easily.

REMEMBER!

In complex arguments like this where there are a multitude of factors to take into account, you may not always be right. It is the thinking that is important.

MATHS SKILLS

Translate information between graphical and numerical form

Learning objectives:

- to represent information from pie charts numerically
- to represent information from graphs numerically
- to represent information from numerical form graphically.

Is it easier to read lists of numbers or to see those numbers visually represented? It often depends on the context. What is important is the ability to change between one form and another and be able to read data tables, histograms and graphs. Can you choose the best scales to construct graphs?

Representing information with pie charts

Some basic chemicals are needed to synthesise other chemicals. How big is that sector of the chemical industry compared with other chemicals that are made? The answer is 22%.

Look at the **pie chart.** What percentage of the industrial chemicals made are used as fertilisers? The answer is 4%.

Is that more or less than the chemicals used for soaps and detergents? The answer is less, as 5% are used for soaps and detergents.

What is the biggest sector of use of chemicals for synthesis? 30% are used in the synthesis of drugs and medicines. These are called pharmaceuticals.

How do you know how big to draw each sector? If the numbers are given as percentages then the pie chart is divided equally into 100 segments. As there are 360° in a circle each segment takes up 3.6°. What size, in degrees, is the paints, pigments and dyes sector? The answer is 36° as the sector represents 10%, so the size is 10 × 3.6 = 36°.

1. What is the percentage of chemicals used to synthesise polymers?

2. How many degrees does the cosmetics and toiletries sector measure?

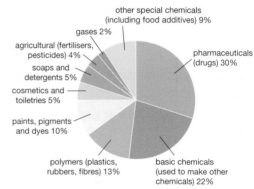

Figure 10.25 Chemical industry

KEY INFORMATION

Can you see that, in this case, the pie chart starts at '12 o'clock' and the sectors are drawn in order of size going clockwise. The first sector is 30%, whereas the second sector represents only 22%. The sectors go all the way round to gases which represents only 2%. Why is the last sector larger? This is because it is an 'other' sector that holds all the smaller contributors together; it is put at the end.

Drawing graphs from data

First the data must be recorded so that the independent variable is measured at regular intervals. The dependent variable is recorded against this.

Once the data has been collected you need to look at the maximum values of the independent variable and choose a **scale** to write along the x-axis, with even steps and values written against each major gridline. For example, if the first value is 0 and the last one is 70 then the scale will be every 10 units to each of the 7 darker gridlines.

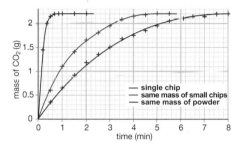

Figure 10.26

In the **graph** in Figure 10.26, the first value is 0 and the last one is 8.

The reaction gave off between 0 and 2.5g of CO_2 so the scale chosen is 0.5 units to every major gridline. The y-axis is labelled 'Mass of CO_2 (g)'.

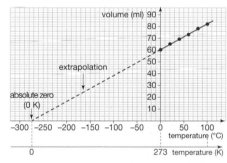

Figure 10.27 Volume of gas against temperature

The graph in Figure 10.27, showing the volume of a gas against the temperature of the gas, has been extrapolated through the negative values of temperature so it is drawn in two quadrants.

3 Draw a pie chart to represent the percentage use of chlorine in other products using this data:

PVC	Other polymers	Disinfectants and cleaners	Solvents	Water treatment	Paper treatment	Other compounds
36%	20%	14%	6%	5%	5%	14%

4 Draw a line graph of this data:

Time in mins	0	2	4	6	8	10
Mass of CO_2 in g	0	0.2	0.38	0.42	0.42	0.42

Check your progress

You should be able to:

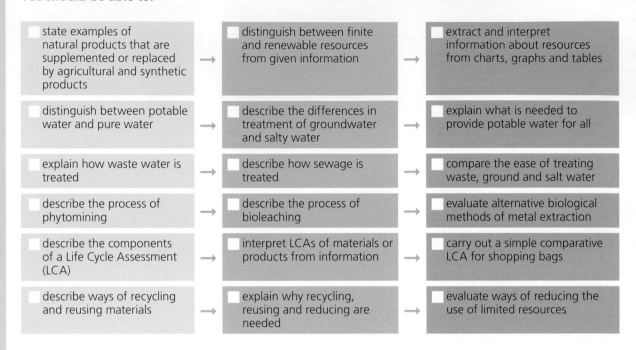

state examples of natural products that are supplemented or replaced by agricultural and synthetic products → distinguish between finite and renewable resources from given information → extract and interpret information about resources from charts, graphs and tables

distinguish between potable water and pure water → describe the differences in treatment of groundwater and salty water → explain what is needed to provide potable water for all

explain how waste water is treated → describe how sewage is treated → compare the ease of treating waste, ground and salt water

describe the process of phytomining → describe the process of bioleaching → evaluate alternative biological methods of metal extraction

describe the components of a Life Cycle Assessment (LCA) → interpret LCAs of materials or products from information → carry out a simple comparative LCA for shopping bags

describe ways of recycling and reusing materials → explain why recycling, reusing and reducing are needed → evaluate ways of reducing the use of limited resources

Worked example

1 **Explain the difference between glass reusing and glass recycling.**

Glass bottles can be collected, washed and refilled to use again. Or they can be taken to a recycling centre, crushed and made into something else.

This answer is correct. It is made clear which is the re-use and which is the recycling by using the words 'recycling centre', but it really needs to be stated which is the reuse and which is the recycling.

2 **What are the four stages of a Life Cycle Assessment, (LCA)?**

extracting, processing, using, disposing

The answer is almost correct. It needs to be made clear whether processing means processing the extracted materials or processing the material to make a set of manufactured goods.

3 **Suggest three ways that the amount of plastic used in packaging could be reduced.**

Use other materials, like paper or fabric. Don't use as much packaging. Provide bins to recycle the plastic.

The student has given three separate ways this could be done. Care needs to be taken not to repeat one suggestion with a very similar one, for example, use cardboard, when paper and fabric have already been suggested.

4 **Suggest, with reasons, whether it is easier to produce clean drinking water from groundwater, from sewage or from the sea.**

I think it is easier to get clean water from the sea as it only needs one process not four processes.

This answer is not complete. The student needs to identify the steps and processes needed for each method of purification. The student then needs to justify why the process chosen is better than the others and to consider the cost of expensive distillation methods compared with holding water in an anaerobic treatment tank.

5 **Explain what 'phytomining' means.**

Plants are used to draw up metals through their roots. The metals stay in the leaves. They are burned and the metal is extracted from the ash.

This answer is correct. A better word to use than 'stay' is accumulate.

End of chapter questions

Getting started

1. Identify the natural product.

   a glass　　　**b** cement　　　**c** wood　　　**d** steel　　　`1 Mark`

2. Which stage is used to destroy microbes when treating water?

   a filtration　　**b** sedimentation　　**c** chlorination　　**d** distillation　　`1 Mark`

3. Humans make use of the Earth's resources for four basic needs. Write down two of these needs.　　`2 Marks`

4. Two mobile phones have been designed. The one that has the more positive Life Cycle Assessment, LCA, is the one that

   a is imported long distances　　**b** is made from plastic
   c has limited battery life　　**d** is carbon neutral in manufacturing　　`1 Mark`

5. Identify which of these resources is not a finite resource.

   a wheat　　**b** crude oil　　**c** coal　　**d** shale gas　　`1 Mark`

6. Two students are concerned about using resources wisely. They discuss whether to make a new object from wood from a managed forest or plastic from crude oil. They choose wood. Suggest why.　　`2 Marks`

7. Match the fertiliser made to the reactants used.

ammonia and phosphoric acid	ammonium sulfate
ammonia and sulfuric acid	potassium nitrate
potassium hydroxide and phosphoric acid	ammonium phosphate
potassium hydroxide and nitric acid	potassium phosphate

   `2 Marks`

Going further

8. Suggest what should happen to waste glass to reduce the use of resources. Explain your answer.　　`1 Mark`

9. State three sterilising agents that are used to produce potable water.　　`1 Mark`

10. Describe two ways that water for processing is stored.　　`2 Marks`

11. A new community is being set up near a river and a forest. Describe four things the community needs or could do to ensure a sustainable existence.　　`4 Marks`

12. Look at Figure 10.12. Identify the sector that was the biggest user of water between 1900 and 2000. After which date was there a surge in water use? Suggest why.　　`2 Marks`

More challenging

13. Suggest how potable water is made from seawater.　　`1 Mark`

14. Suggest why water is sterilised with chlorine before being distributed to customers.　　`1 Mark`

15. Describe the steps used to process sewage.　　`2 Marks`

Most demanding

16 Describe four comparisons for an LCA that you would make in deciding whether to use plastic-lined milk cartons or glass bottles in your new company.

4 Marks

17 Explain the steps needed to extract metals using phytomining.

4 Marks

18 Explain the process of bioleaching.

2 Marks

19 Explain how copper is purified from solutions made from alternative methods of extraction.

2 Marks

20 Copper can be extracted by smelting or by other methods. Several factors are taken into consideration when deciding on the method to use, as in the table below.

4 Marks

	Smelting	Alternative method
Percentage extracted	60%	90%
Cost of capital equipment	£600 000	£300 000
Operational cost	£3.20/kg	£3/kg
Time taken per unit	36 hours	144 hours
Emission of CO_2	0.6 kg per 1 kg of Cu	0
Emission of SO_2	1 kg per 1 kg of Cu	0

Evaluate the data provided to suggest which should be the preferred method and suggest another factor for consideration that is missing from the table.

Total: 40 Marks

Key

relative atomic mass
atomic symbol
name
atomic (proton) number

1	2											3	4	5	6	7	0
																	4 **He** helium 2
7 **Li** lithium 3	9 **Be** beryllium 4											11 **B** boron 5	12 **C** carbon 6	14 **N** nitrogen 7	16 **O** oxygen 8	19 **F** fluorine 9	20 **Ne** neon 10
23 **Na** sodium 11	24 **Mg** magnesium 12											27 **Al** aluminum 13	28 **Si** silicon 14	31 **P** phosphorus 15	32 **S** sulfur 16	35.5 **Cl** chlorine 17	40 **Ar** argon 18
39 **K** potassium 19	40 **Ca** calcium 20	45 **Sc** scandium 21	48 **Ti** titanium 22	51 **V** vanadium 23	52 **Cr** chromium 24	55 **Mn** manganese 25	56 **Fe** iron 26	59 **Co** cobalt 27	59 **Ni** nickel 28	63.5 **Cu** copper 29	65 **Zn** zinc 30	70 **Ga** gallium 31	73 **Ge** germanium 32	75 **As** arsenic 33	79 **Se** selenium 34	80 **Br** bromine 35	84 **Kr** krypton 36
85 **Rb** rubidium 37	88 **Sr** strontium 38	89 **Y** yttrium 39	91 **Zr** zirconium 40	93 **Nb** niobium 41	96 **Mo** molybdenum 42	[98] **Tc** technetium 43	101 **Ru** ruthenium 44	103 **Rh** rhodium 45	106 **Pd** palladium 46	108 **Ag** silver 47	112 **Cd** cadmium 48	115 **In** indium 49	119 **Sn** tin 50	122 **Sb** antimony 51	128 **Te** tellurium 52	127 **I** iodine 53	131 **Xe** xenon 54
133 **Cs** cesium 55	137 **Ba** barium 56	139 **La*** lanthanum 57	178 **Hf** hafnium 72	181 **Ta** tantalum 73	184 **W** tungsten 74	186 **Re** rhenium 75	190 **Os** osmium 76	192 **Ir** iridium 77	195 **Pt** platinum 78	197 **Au** gold 79	201 **Hg** mercury 80	204 **Tl** thallium 81	207 **Pb** lead 82	209 **Bi** bismuth 83	[209] **Po** polonium 84	[210] **At** astatine 85	[222] **Rn** radon 86
[223] **Fr** francium 87	[226] **Ra** radium 88	[227] **Ac*** actinium 89	[261] **Rf** rutherfordium 104	[262] **Db** dubnium 105	[266] **Sg** seaborgium 106	[264] **Bh** bohrium 107	[277] **Hs** hassium 108	[268] **Mt** meitnerium 109	[271] **Ds** darmstadtium 110	[272] **Rg** roentgenium 111							

1 **H** hydrogen 1

Elements with atomic numbers 112–116 have been reported but not fully authenticated

* The Lanthanides (atomic numbers 58–71) and the Actinides (atomic numbers 90–103) have been omitted.
Relative atomic masses for Cu and Cl have not been rounded to the nearest whole number.

Glossary

A

acids dissolve in water to produce solutions with a pH of less than 7

acid rain rain which has been made more acidic by pollutant gases

activation energy the energy needed for a chemical reaction to happen

addition polymer a very long molecule resulting from polymerisation, e.g. polythene

aggregate gravel added to cement and sand to make concrete

alkali metals the metals in Group 1 of the periodic table

alkalis compounds which produce hydroxide ions in water

alkanes a family of hydrocarbons with all single carbon-carbon covalent bonds and general formula C_nH_{2n+2}

alkenes a family of hydrocarbons with one double carbon-carbon bond and general formula C_nH_{2n}

allotropes different forms of the same element

alloy a mixture of a metal with one or more other metals or non-metals to change the properties of the metal

alpha particles radioactive particles which are helium nuclei – helium atoms without the electrons (they have a positive charge)

ammeter meter used in an electric circuit for measuring current

anion ion with a negative charge; they move to the anode during electrolysis

anode electrode in electrolysis with a positive charge

aquifer underground layer of permeable rock or loose materials (gravel or silt) where groundwater is stored

atom the basic 'building block' of an element, the smallest part of an element that can take part in a chemical reaction

atomic number the number of protons in the nucleus of an atom

Avogadro's constant the number of atoms, molecules or ions in one mole of a given substance, and is 6.02×10^{23} per mole

B

balanced symbol equation chemical equation written in chemical symbols showing the number of atoms on each side of the equation balance

base reacts with an acid to form a salt

bioleaching process that uses bacteria to leach metal compounds from rocks

biological catalyst molecules in cells of living organisms that speed up chemical reactions

boiling point temperature at which the bulk of a liquid turns to vapour

buckminsterfullerene a very stable spherical structure of 60 carbon atoms joined by covalent bonds (an allotrope of carbon)

C

carbon an element present in all living things and forms a huge range of compounds with other elements

carbon-14 radioactive isotope of carbon

carbon dioxide (CO_2) a greenhouse gas which is emitted into the atmosphere as a product of combustion

carbon footprint the total amount of carbon dioxide and other greenhouse gases emitted over the full life cycle of a product, service or event.

catalyst a chemical that speeds up a reaction but is not used up by the reaction

cathode the negative electrode in electrolysis

charge(s) a property of matter, charge exists in two forms, positive and negative, which attract each other

chemical properties the characteristic chemical reactions of substances

chlorination addition of chlorine to water supplies to kill micro-organisms

chromatography a method for separating substances, used to identify compounds and check for purity

close-packed atoms structure of many metals

collision frequency the number of collisions between particles that happen in one unit of time

combustion exothermic reaction of a substance with oxygen

compound two or more elements which are chemically joined together, e.g. H_2O

concentration the amount of chemical dissolved in a certain volume of solution

conductors materials which transfer thermal energy easily; electrical conductors allow electricity to flow through them

conservation of energy principle stating that energy cannot be created or destroyed

conservation of mass the total mass of reactants equals the total mass of products formed in a chemical reaction

covalent bonds bonds between atoms where a pair of electrons is shared

cracking the process of breaking down large hydrocarbons into smaller molecules

curved line line of changing gradient

D

decay to rot or decompose

delocalised electrons electrons which are free to move from atom to atom in a giant structure or a molecule

density the density of a substance is its mass divided by its volume

diesel oil fuel for diesel engines, traditionally obtained from crude oil but other forms such as biodiesel have been developed

direct current an electric current that flows in one direction only

displacement reaction chemical reaction where an element takes the place of or 'pushes out' another element from a compound

distillation the process of evaporation followed by condensation

dot and cross diagram a diagram representing the number of electrons in the outer shell of atoms or ions

E

electrical conductivity a measurement of the ability to conduct electricity

electrical conductors materials that let electricity pass through them

electrode ions are discharged at the electrodes during electrolysis

electrolysis the process of passing direct current through a melted ionic compound or a solution of an ionic compound so ions are discharged and the compound is broken down

electrolyte a liquid or solution that conducts electricity and breaks down during electrolysis

electronic structure the arrangement of electrons in the sequence that they occupy the shells or energy levels, e.g. the 11 electrons of sodium are arranged 2,8,1

electrons small negatively charged particles within an atom that are outside the nucleus

electrostatic attraction attraction between opposite charges, e.g. between Na^+ and Cl^-

elements substances made out of only one type of atom with the same number of protons in the nucleus

empirical formula simplest ratio of atoms or ions in a compound

endothermic reaction chemical reaction which takes in thermal energy

energy the ability to 'do work '

enzymes biological catalysts that increase the speed of chemical reactions

equilibrium when the forwards and backwards reactions are occurring at the same rate in a closed system

estimate calculate approximately the value of something

evaporation when a liquid changes to a gas, it evaporates

exhaust gases gases discharged into the atmosphere from an engine as a result of combustion of fuels

exothermic reaction chemical reaction in which thermal energy is given out

explosion a sudden, loud, violent release of energy by a chemical reaction

extrapolation making an estimate by continuing a trend or graph line beyond the range of results

F

filtration the process of using a porous material to remove solids from water or solutions

formulation a mixture that has been designed as a useful product

fractional distillation crude oil is separated into fractions using this process of distillation where a mixture of liquids is vaporised and compounds with different boiling points condense at different temperatures

force a push or pull which is able to change the velocity or shape of a body

fossil fuels fuels which are the fossilised remains of plants or animals, such as coal, oil and gas

fullerenes cage-like carbon molecules containing many carbon atoms, e.g. buckyballs

G

giant covalent structure a large regular arrangement of atoms all joined together by covalent bonds

giant ionic lattice the regular three-dimensional arrangement of ions in an ionic compound , also called a giant ionic structure

gradient rate of change of two quantities on a graph; change in y divided by change in x

graphite a type of carbon made of layers of atoms

greenhouse gas any of the gases whose absorption of solar radiation is responsible for the greenhouse effect, e.g. carbon dioxide, methane

group within the periodic table the vertical columns are called groups

Group 1 the elements in Group 1 of the periodic table, the alkali metals

Group 7 the elements in Group 7 of the periodic table, the halogens

H

half equation a redox reaction is made up of two half equations, one in which electrons are lost and one in which electrons are gained.

halogens reactive non-metals in Group 7 of the periodic table, e.g. chlorine

hardness resistance of a solid material to cutting, indentation or scratching

homologous series a series of organic compounds that have the same general formula, i.e. the general formula of alkanes is C_nH_{2n+2}

hydrocarbons compounds containing only hydrogen and carbon

I

incomplete combustion takes place when there is not enough oxygen present for complete combustion

indicator used to show pH of a solution or when the end point of a titration is reached

insoluble salt salt which is not soluble in water

intermolecular force force between molecules

interpolation making an estimate of a value from values on either side of the point

ionic bond the chemical bond between ions of opposite charges

ionic equation an equation showing changes to the ions involved in a reaction

ionises adds or removes electrons from an atom leaving it charged

ions charged particles (can be positive or negative)

isotopes atoms with the same number of protons but different numbers of neutrons

J

joule unit of work done and energy

K

kilogram (kg) unit of mass

kinetic energy the energy that moving objects have

L

Le Châtelier's principle if a system is at equilibrium and a change is made to any of the conditions, then the system responds to counteract the change

life cycle assessments (LCAs) are carried out to assess the environmental impact of products in each of the stages involved in their manufacture, use and disposal

limewater a solution of calcium hydroxide in water – the colourless solution turns milky in the presence of carbon dioxide

limiting reactant chemical used up in a reaction that limits the amount of product formed

line spectrum a spectrum produced by gaseous atoms showing individual lines at particular wavelengths that is unique for each element

lustrous shiny

lysis to split apart

M

magnitude size of something

mass the amount of matter in something; it is measured in kilograms (kg)

mass number the sum of the number of protons and neutrons in a nucleus

melting point the temperature at which a solid turns into a liquid

metal halide a compound of a halogen and a metal, e.g. potassium bromide

metallic bonding the bonding between atoms in a metal due to delocalised electrons

metallic properties the physical and chemical properties specific to a metal, such as lustre, electrical conductivity and the ability to form positive ions

metalloids elements with properties of both metals and non-metals; in the periodic table they are between the metals and non-metals

metals elements that are usually solid, lustrous, conduct electricity and form ions by losing electrons

minerals natural solid materials with a fixed chemical composition and structure, rocks are made of collections of minerals

mobile phase in chromatography this is the phase that moves

mole a unit for a standard amount of a substance. One mole of any substance contains the same number of particles, atoms, molecules or ions as one mole of any other substance

molecular formula the formula of a chemical using chemical symbols, e.g. methane has the molecular formula CH_4

molecule two or more atoms covalently bonded to form the smallest unit of an element or compound, e.g. O_2, H_2O

molten a substance in its liquid state, often referring to a substance which is solid at ordinary temperatures, such as rock, ores, metals or salts, when heated to temperatures above its melting point

N

negative ion an ion with a negative charge, such as when atoms gain electrons

neutral a neutral solution has a pH of 7

neutralisation the reaction that takes place when an acid and base react to produce a salt and water

neutron particle which does not have a charge found in the nucleus of an atom

non-metals elements that are solids, liquids or gases that do not conduct electricity and bond covalently or form negative ions by their atoms gaining electrons

non-renewable something which is used up at a faster rate than it can be replaced e.g. fossil fuels

nucleus central part of an atom that contains protons and neutrons

O

optimum conditions the conditions, such as temperature and pressure, that give the products of a chemical process at the lowest cost

order of magnitude values that differ by one order of magnitude are 10 times larger or smaller than each other

oxidation when a reactant gains oxygen or loses electrons

P

particulates small particles in the air often caused by burning fuels

period a row in the periodic table

periodic table a table of all the chemical elements in order of their atomic numbers

petrol volatile mixture of mainly hydrocarbons used as a fuel

pharmaceuticals medical drugs

physical property property that can be measured without changing the chemical composition of a substance, e.g. hardness

phytomining process that uses plants to extract metals

pollutants substances that can cause damage to the environment

pollute put unwanted or harmful substances into the environment

pollution contamination of the environment as a result of human activities

polymer very large molecule formed from many similar smaller molecules (monomers) linked together

positive ion an ion with a positive charge, such as when atoms lose electrons

potable water water that is safe to drink

precipitate solid formed in a solution by a chemical reaction

precipitation reaction chemical reaction in which a solid is formed when two solutions are mixed, e.g. in chemical tests for ions

product substance produced by a chemical reaction (shown on the right-hand side of the chemical equation)

protons positively charged particles found in the nucleus of an atom

pure a pure substance is a single element or compound that is not mixed with any other substance

R

random having no regular pattern

rate of reaction the speed with which a chemical reaction takes place, measured by the amount of a reactant used or amount of product formed in a given time

reactants chemicals that react together in a chemical reaction (shown on the left-hand side of the chemical equation)

recharging battery or cell being charged with a flow of electric current

reduction when a reactant loses oxygen or gains electrons

refine the refining process turns crude oil into usable forms such as petrol

relative atomic mass the mass of an atom compared to 1/12 of the mass of a carbon-12 atom

relative formula mass the sum of the relative atomic masses in a compound

renewable energy energy from a resource that is rapidly replaced

renewable resource any resource that can be replenished at the same rate that it is used, e.g. biofuels

reservoir a water resource where large volumes of water are held

reversible reaction a chemical reaction where the reactants form products that, in turn, react together to give the reactants back

R_f in chromatography is the distance a substance moved divided by the distance the solvent moved

S

saturated hydrocarbon a hydrocarbon containing the maximum number of hydrogen atoms and only single carbon-carbon bonds; alkanes are saturated hydrocarbons

sea water water from the sea that contains high levels of dissolved salts making it undrinkable

sedimentation a process during water purification where small solid particles are allowed to settle

single covalent bond chemical bond between atoms where each atom shares one pair of electrons

solar energy energy from the Sun

soluble a soluble substance can dissolve in a liquid, e.g. sugar is soluble in water

solution when a solute dissolves in a solvent, a solution forms

solvent the liquid used to dissolve a solute

stable electronic structure the electronic structure of a noble gas, with two electrons in the first shell and eight electrons in every other outer shell, e.g. He 2; Ne 2,8; Ar 2,8,8; Kr 2,8,18,8

standard form a way of writing a large number with one number before the decimal point, multiplied by a power of 10, e.g. 1 200 = 1.2×10^3

stationary phase the phase in chromatography that does not move; in paper chromatography it is the paper

straight line line of constant gradient

strength (of an acid) strong acids ionise completely in water; weak acids partially ionise

sub-atomic particles particles that make up an atom, e.g. protons, neutrons and electrons

sublimation change of state of a substance from a solid directly to a gas; e.g. iodine

T

thermal decomposition the breaking down of a compound into two or more products on heating

thermal energy energy that can be transferred as heat

toxic a toxic substance is one which is poisonous and causes harm to living organisms

transition element an element in the middle section of the periodic table, between the block containing Groups 1 and 2 and the block containing Group 3 to Group 0

U

unsaturated hydrocarbon a hydrocarbon containing fewer than the maximum number of hydrogen atoms possible, and so at least one double bond.

V

vacuum space containing no particles of matter

voltage (also called the potential difference) the difference in electrical potential between two points or objects

voltmeter instrument used to measure voltage (potential difference)

volt (V) unit used to measure voltage

W

water conservation reducing water consumption through planned choice, e.g. hosepipe bans and water metering

water resources places from where water is extracted or where it is stored, e.g. aquifers, reservoirs or lakes

wavelength distance between two wave peaks or the distance between identical points in adjacent cycles of a wave

Index